"ONCE YOU SEE THE STRINGS, THE PUPPET SHOW IS NEVER THE SAME."

UNFOLLOW: Escaping The Distraction Machine

© 2025 ALT LANE JOURNALS

First Edition

ISBN: 978-1-7642383-3-5
Published by ALT LANE JOURNALS
Sydney, Australia

Disclaimer:

This book is intended for personal growth and awareness-building purposes only. It is not a substitute for professional advice, diagnosis, or treatment by a qualified medical, mental health, or counselling professional. Readers who feel overwhelmed, anxious, or distressed by their social media or technology use should seek support from an appropriate professional or trusted resource.

UNFOLLOW: ESCAPING THE DISTRACTION MACHINE

ALT.LANE

"YOUR ATTENTION IS BEING BOUGHT, SOLD, AND STOLEN AT AN INDUSTRIAL SCALE! WHAT WILL YOU DO TO TAKE IT BACK?"

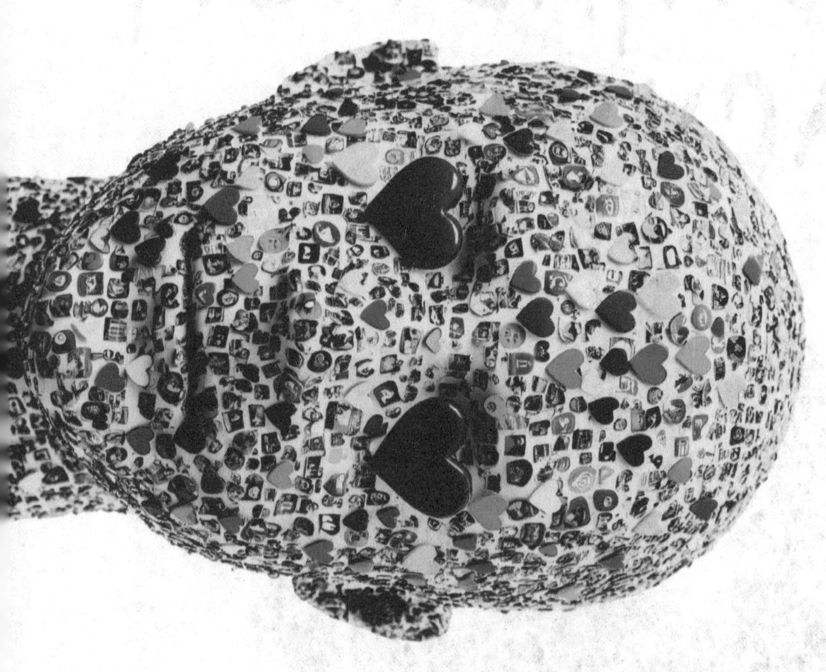

ALT.LANE

Introduction:
The Illusion of Connection

You scroll. You like. You react. You post.

Somewhere between the third video and the eighth suggested reel, a whole hour has disappeared. You don't remember what you saw. You don't feel better than before. You didn't come here for this, but here you are again. This is not your fault.

This is the machine doing exactly what it was built to do: distract you. It's fast. It's endless, and it's everywhere.

Social media has become the central nervous system of modern life—plugged into our palms, pulsing through every interaction, shaping how we think, feel, relate, and exist.

Yet, for something so embedded in our daily routines, few of us ever stop to ask: At what cost?

The Trap We Didn't See Coming

No one saw it coming. Not like this.

What began as a new way to stay in touch with friends, to share photos, to connect across borders. It felt exciting—liberating—human.

To be fair, the early pioneers of social media likely didn't intend for things to end up where they are now. They were innovators, not villains. They didn't set out to build a system that would erode mental health, fracture society, and monopolise attention. They were chasing curiosity, not control.

But that's the thing about tools—they evolve based on how they're used, and have done so through the course of humanity.

And social media evolved fast.

Once companies realised that human attention could be monetised, the game changed. Likes became currency. Shares became influence. Algorithms became the new gods.

Now, a new generation of tech creators stand on the shoulders of those early pioneers—not to build connection, but to manufacture dependency. Their goal isn't to help you communicate. It's to keep you hooked long enough to sell you, because the longer you scroll, the more they earn.

So, what began as a promising tool for connection has become a global infrastructure of distraction, designed to pull you away from your present moment —and keep you there.

It didn't start as a trap, but that's where we've ended up. And no one is coming to fix it.

Distorted Reflections

If you want to understand what social media is doing to us, next time you open an app, view it for what it is. Scroll through TikTok. Scroll through Instagram. Scroll through Snapchat stories. Watch what people are becoming.

We live in a world where influence is a career, where Likes are currency, and where shock equals success. The algorithm rewards whatever generates the most engagement—so what rises to the top is rarely wisdom, compassion, or creativity. It's ego, exposure, and extremes.

We've built a system where validation is addictive—and the most addictive validation is visual. So, we perform. We pose. We push boundaries.

- Selfies in hospital beds after cosmetic surgery. ("LOVE, I need this!!!!!!")

- Teens lip-syncing to explicit lyrics in hyper-sexualized poses. ("OMG queen slay all day")

- Dangerous stunts staged for views. ("Bro's got no fear—respect")

Influencer culture has become a stage for the performance of identity, where authenticity is filtered, curated, and sacrificed for engagement.

Even worse, children are being trained to believe that how they look, how much they're seen, and how well they perform for an algorithm determines their value. And when that's the system, what kind of behaviours rise to the surface?

The outrageous. The unfiltered. The sexual. The dangerous. Because in the attention economy, dignity is a liability.

It's not just toxic—it's **sad**.

Yet, this is the version of reality billions of people are scrolling through, every single day.

So, the question becomes: What kind of world are we building—when we're all performing for the machine?

It's Not Just You. It's Everyone.

This book is not here to shame you. You're not weak. You're not lazy. You're not broken. Everyone is in the same place.

Go to any airport, any train station – go to any festival, any concert – drive past any school bus stop… it's glaringly obvious that there is a massive issue at hand.

We are now living in a world that's being deliberately designed to weaken our willpower, fragment our presence, and keep us addicted to a feed that feeds on us.

What was once a family dinner is now four people staring at four different devices. What was once a moment of silence first thing in the morning, is now a reflex to open an app. What was once a teenager's inner life is now an algorithm's target audience. And what was once a home—a safe and private space—now a fishing net of microphones and cameras, catching everything—always listening, always watching.

Why This Book Exists

This book has been written, because we all already know something is very wrong.

You've felt it—the hollow feeling inside after another pointless, empty scrolling session, where an hour has passed and you're snapping back to reality, thinking "what am I doing"!

You've noticed the anxiety that creeps in when your phone isn't in your hand. That moment where you have a free minute in your evening, but you must, MUST, MUST, fill that minute by checking your feed.

You've seen your kids disappear into screens, growing quieter, quicker to react, harder to reach. Giggling to themselves whilst scrolling, whilst you observe, wondering "what are you looking at?" And deep down, you're thinking ... What would life look like without this in today's times?

This book is here to explore that question. An attempt at a wake up call for humanity—to see the system for what it is, and to ask:

What is it costing me?

What is it costing us?

AND, do I accept?

What You'll Discover

In the pages that follow, we'll expose the mechanics behind social media's grip:

- The behavioural design used to hook you.
- The psychological toll of being always online.
- The data economy built on your identity.
- The way it's reshaping childhood, family, and thought itself.
- And most importantly—how to get out.

Because escaping the distraction machine isn't just about logging off, it's about taking back your mind, your time, and your humanity. It's about unfollowing everything that keeps you connected to something that wastes your time.

We weren't born for this. But it's also not too late to live differently.

So, let's begin.

Contents

Unfollow: Escaping The Distraction Machine

Unfollow In Action: Your Social Media Escape Blueprint

Chapter 1:
The Distraction Economy

There is a war happening every time you unlock your phone. It's not a war for land, oil, or power. It's a war for something far more valuable: your attention.

And right now, you're losing.

Attention is Currency

For most of history, economies were built on physical resources—gold, oil, steel, labour. But in only the last 10 years, the most valuable commodity on Earth is no longer something you can hold in your hands.

It's what you give away every time you look at a screen: your attention.

Every second you spend online is worth money to someone. That TikTok—That Instagram scroll—That YouTube autoplay—Each second is monetised, tracked, measured, and sold.

And it doesn't stop with passive viewing. Once you become a creator—even casually—you're no longer just participating in the system. You're feeding it.

Every post you make becomes a mini performance, measured and critiqued by a matrix of numbers called Post Insights.

- Views
- Likes
- Comments
- Shares
- Saves
- Reach
- Profile clicks
- Follows

Each metric tells a story and each number carries weight.

- A post with high views but low likes? Maybe your content wasn't engaging enough.
- Lots of likes but no new followers? Maybe you didn't show enough value.
- High reach but low saves? Maybe it was shallow.
- Plenty of comments, but mostly emojis? Not real engagement.

This isn't casual anymore. It's evaluation. By your peers, by strangers, and by the algorithm itself.

The user starts asking themselves:
- "How did this post perform?"
- "What does that say about me?"

And the platform asks:
- "How can we push this further?"

The algorithm doesn't just analyse your content—it analyses how others respond to it, and then uses that data to determine what gets promoted next.

The more you check your insights, the more you train yourself to value your content not by meaning, but by performance.

And the more your identity becomes entangled with metrics, the more you become a puppet for the machine!

You're not just a user. You're the product being packaged and auctioned off to the highest bidder, and the tech giants aren't just bystanders in this economy—they are the enablers.

Human Lives, Measured in Metrics

It's easy to think of this in abstract terms—algorithms, platforms, engagement. But this economy is built on human lives, broken into data points.

Every second you're online is being:

- Tracked by pixels
- Scored by algorithms
- Monetised by corporations
- Sold to advertisers

When you're not "checking your feed," you're not feeding the system—with your behaviour, your emotions, and your future.

In this economy, a company's success is measured by how long they can keep your eyes on a screen. Your distraction is their profit margin, so therefore your addiction is their business model.

The longer you stay, the more valuable you become—to everyone but yourself.

The Birth of the Feed

The moment everything changed was the invention of the infinite feed—a scroll that never ends. What started as a convenience quickly became a poker machine for the brain.

Every time you scroll, your brain receives a tiny dose of dopamine—the same chemical activated by gambling, sugar, and addictive drugs.

That moment of "what's next?" is deeply satisfying for some reason. But here's the trap: The more you scroll, the more they learn about you. The more they learn about you, the more they can predict you. And the more predictable you are, the easier it is to keep you hooked.

This isn't accidental. It's the result of billion-dollar behavioural engineering.

The Economy of Distraction

Social media doesn't make money when you're fulfilled, inspired, or focused. It makes money when you're hooked, distracted, and reactive.

Platforms like Instagram, YouTube, TikTok, and Snapchat have one mission:

maximise time-on-platform. And to do that, they must keep you:

- Clicking instead of thinking
- Reacting instead of reflecting
- Scrolling instead of stopping

Distraction isn't a side effect, it's the product. Every minute you spend online is carefully designed to prevent you from doing anything else—working, resting, spending time with your family, being bored, even thinking.

Because if you stop, you might leave… and if you leave, they stop making money.

The Myth of Multitasking

Some people defend their screen time with the myth of multitasking. "I'm productive." "I'm just unwinding." "I'm keeping up." But what they are really doing is dividing their presence across dozens of fragments, never fully investing in any one thing.

Social media doesn't want your full attention—it wants your constant attention. You don't have to be thinking deeply. You just have to be checking in.

A like here. A reply there. A reaction. A view. A swipe.

This isn't harmless fun. It's changing your brain, and over time your ability to focus, to read deeply, to sit with a thought without distraction—all of it erodes.

The muscle of attention atrophies when you constantly outsource your engagement to an algorithm. And that algorithm is training you—whether you realise it or not—to never stay in one place for long periods of time.

According to Facebook/Meta internal research from 2021, the average user spends just 1.7 seconds looking at a piece of content on platforms like TikTok and Instagram Reels. For younger users, that number is often under 1.5 seconds.

That means in a single minute of scrolling, you've consumed over 30 distinct pieces of content—images, captions, videos, audio, text, sometimes all at once and then it's gone. Swiped past. Forgotten.

What does that do to a mind? It creates a state of hyper-stimulation and shallow focus, where nothing truly sinks in. Multitasking in this environment isn't just ineffective—it's destructive, because you're not truly doing many things at once. You're doing everything at half-depth.

In the process, your brain forgets how to slow down, how to concentrate, how to stay with something long enough to understand it fully.

It's a slow, systematic erosion of your cognitive depth—and the world gets noisier every time it happens.

Why This Matters

Your attention is more than time, it's your life force. Whatever owns your attention, shapes your reality. It determines what you believe, how you feel, who you trust, and what kind of world you think you live in.

Right now, your attention is being bought, sold, and manipulated at industrial scale and every time you look away from your real life to check your feed, someone else is profiting. Not from your happiness. Not from your growth.

But from your distraction.

What Is Your Time Worth?

If attention is the new oil, then you are the well—constantly drilled, extracted, and emptied.

So, the question isn't just "how much time are you spending online?" The question is: "What are you giving up every time you scroll?"

Because while the world races to capture more of your attention, you only get one life… and that one life is slipping away second by second, right now— whether you're paying attention or not.

ALT.LANE

Chapter 2:
The Architecture of Addiction

Most people don't realise they're addicted to social media. That's the genius of it. The trap isn't obvious. It doesn't scream. It whispers. It rewards you in small doses, just enough to keep you coming back.

This is not an accident. It's design.

Behind every scroll, like, streak, and notification, there's an entire industry of engineers, behavioural psychologists, and machine learning experts working to hijack your brain's reward system.

And they've done it brilliantly.

Designed to Hook You

You may think you use social media because it's fun or useful. But the truth is, you use it because it was built to be irresistible. Tech companies don't guess what you'll click on next—they predict it. They don't hope you'll keep scrolling—they design the experience so you can't stop. This is the architecture of addiction—a system so effective, even its creators aren't immune.

Former executives from multiple social media platforms have admitted publicly that they don't let their kids use the products they helped build. Why? Because they know what you're about to learn: These platforms aren't designed for communication. They're designed for compulsion and addiction.

Let's break down how they do it.

1. Infinite Scroll

You never reach the bottom of your feed. There's no natural stopping point. This taps into a psychological effect called "the variable reward loop"—a behavioural design pattern that leverages the human brain's tendency to seek out rewards, and the release of feel-good chemicals in our brain when those rewards are unpredictable. By introducing variability in rewards, it can make activities more engaging and habit forming.

Sometimes you get a post you like. Sometimes you don't. But that, maybe, is what keeps you scrolling... Excited to see what comes next.

2. The Like Button

It started as a simple feature. Just a little thumbs-up. A heart. A double tap.

Facebook introduced the Like button in 2009. It was meant to be a lightweight way for users to interact with content without leaving a comment. Something casual. Frictionless. But almost immediately, it became something much bigger.

People didn't just notice how many Likes they got—they started counting them. It wasn't long before Likes became the new scoreboard of social currency. "How many Likes did you get?" A question once asked as a novelty quickly turned into a measuring stick for worth, relevance, and acceptance.

We then began to post not for the sake of sharing, but for the performance of being seen. Content shifted from spontaneous and real, to curated and strategic.

It changed everything: The way we dressed, the places we went, the food we ate, the way we posed, edited, filtered, captioned, and smiled. Likes turned moments into marketing, users into performers, and life into content.

Once that dynamic took hold, there was no going back. The Like button taught an entire generation that external validation is the ultimate goal, and if you're not being seen or acknowledged—you're doing something wrong.

Even worse, the absence of Likes became a form of punishment. Didn't get enough? The post is deleted. Hidden. Rewritten. You're not enough. That's the silent message we started to internalise.

Social media stopped being about expression. It became about performance, and the Like button was the applause we never knew we were chasing.

3. Notifications

Those alerts that pop up are not innocent nudges. They are carefully engineered triggers, timed and tested to maximise re-engagement.

Their purpose? To pull you back in—even when you had no intention of returning. Every notification is deliberate:

- The colour red—used because it signals urgency, like a stop sign or a warning light.
- The phrasing—"Someone mentioned you in a post," "Someone commented on your reel."
- The timing—sent during your most likely idle hours, when you're between tasks or looking for distraction.

And a phone is the perfect place for them to pop up, as we have conditioned ourselves, not to have our phones away in a bag, or left in the car. We have them next to us, in front of us, all the time.

They glow and ignite with energy when a push notification is received, and if they happen to be in our pocket, no problem, they vibrate so we are aware.

It's all about hijacking your attention—and once your concentration is broken, the system knows it has a shot.

In a 2008 study, conducted by the University of California, research showed that after being interrupted, it can take up to 23 minutes to fully regain focus. That's 23 minutes that we are less productive when we allow ourselves to be continually interrupted.

Every time your phone lights up or buzzes, you don't just lose a moment—you lose the flow. You lose the thought. You lose the rhythm—and if that buzz leads you back to the app, the loop begins again. It's such an effective tool for hijacking attention; other industries adopted it.

News apps now send push alerts with sensationalised headlines—not because the story is urgent, but because clicks mean revenue.

- "New Deadly Virus Detected: Experts Say It May Already Be Spreading."
- "Interest Rates Set to Surge? Economists Warn of a Financial Shock Ahead."
- "Is Your Tap Water Toxic? New Report Claims Millions at Risk — Here's What You Need to Know."

It's not about informing you. It's about interrupting you—because the more eyes on the article, the more ad views they get, and the more profit they make. Just like social media, these platforms don't care about your focus. They care about your engagement.

Notifications are no longer about what you need to know. They're about what someone else profits from you knowing—right now, immediately, before your attention goes elsewhere.

4. Autoplay & Algorithmic Feeds

You don't choose what to watch next. The algorithm chooses for you. It knows your history, your habits, your emotional highs and lows. It serves you content not to inform you—but to keep you on the platform.

Even outrage is useful—if it keeps you scrolling.

Hijacking the Brain

This system works because it speaks the language of your biology. Your brain is wired to seek pleasure, avoid pain, and conserve energy.

Social media taps into all three:

- Pleasure: Dopamine hits from likes, shares, and messages
- Pain avoidance: FOMO, invites, or inside jokes
- Efficiency: Easy, low-effort engagement that replaces real interaction

The result is a feedback loop that rewires your brain. Over time, you train yourself to reach for your phone in every spare moment.

- Bored? Scroll
- Tired? Scroll
- Lonely? Scroll
- Happy? Post
- Sad? Scroll

The device becomes your default comfort, your emotional regulator, your distraction, your entertainment, your identity.

The Addiction No One Admits To

Unlike traditional addictions, social media addiction is not stigmatised. It's normalised, encouraged, and monetised.

You don't hide it—you joke about it. You don't get help for it—you get updates, and the consequences are easier to ignore—until they aren't.

- Sleep loss
- Reduced attention span
- Mental fatigue
- Constant anxiety
- The inability to be present

We don't notice the damage because it's subtle. It accumulates slowly. But by the time we realise we're caught in the loop, the loop has become an integral part our life.

The Profitable Truth

Here's the bottom line: Your addiction is someone else's business model. If you felt content, centered, focused, and fulfilled, you wouldn't need to scroll.

That's why the platforms are built to do the opposite.

- They don't want you satisfied. They want you seeking.
- They don't want your peace. They want your patterns.
- They don't want you to log off, because as long as you're on the app, you're feeding the machine.

Who's in Control?

You didn't choose this addiction, but you can choose to see it for what it is. Because once you see the apps for what they are, you can stop blaming yourself and start asking the right questions: If I didn't build the habit… who did? And, like any other bad habit, what steps can I take to break it?

Chapter 3:
Surveillance as a Feature, Not a Bug

You used to have to break the law to be surveilled. Now, you just have to download an app. And the wildest part?

You agreed to it.

You're Not Being Watched—You're Being Analysed

When we think of surveillance, we imagine someone watching us—security cameras, hackers, government agencies. However, that's outdated.

Surveillance today doesn't look like someone spying on you. It looks like an app asking for your location. It looks like agreeing to "Allow All" on a cookie banner. It looks like clicking "I Accept" without reading, and…

- We do it constantly
- Casually
- Without hesitation
- Without reflection

It's one of the most significant and normalised violations of privacy in modern life, and we've been trained to believe it's just part of the process.

- Click the box, move forward
- Agree to the terms, get to the content
- No checkbox, no access

Here's the problem though: Every time we blindly click "Accept All," we surrender more than just convenience.

We're giving up:

- Our location
- Our behavioural patterns
- Our voice, camera, and contacts
- Our digital footprint—and everything it reveals about who we are.

This has become the modern trade-off. Access in exchange for exposure. Freedom in exchange for surveillance—and we've stopped questioning it.

The interfaces are designed to wear us down. Too many steps, too much legal jargon, and the quickest option—the one that benefits them most—is made the easiest to click. We've normalised this behaviour so deeply that it's no longer even considered a decision. It's a reflex. But this reflex has consequences.

It sets a precedent:

- Your privacy is negotiable
- Your autonomy is optional
- Your attention is for sale

Once a culture accepts that, it becomes much easier for the next platform, the next device, the next policy—to take even more, because no one's resisting. We're too busy trying to get to the next page.

Modern surveillance isn't just about observation. It's about harvesting our behaviour, mapping our psychology, and predicting our future decisions—so they can be monetised.

And it's happening across every platform you use:

Snapchat

Despite its branding around disappearing messages, making it seem like its non intrusive towards personal information, Snapchat has come under fire for using tools like Snap Map, which tracks your exact GPS location in realtime—even when the app isn't actively open.

Users can opt into "Ghost Mode," but even then, the app continues to collect location data in the background.

Snapchat also stores metadata about your Snaps—who you send them to, when, and how often—feeding a behavioural profile more permanent than the content itself.

Instagram

Instagram doesn't just track what you like. It monitors, what you hover over, how long you watch a reel, what hashtags you follow, whose stories you skip or rewatch.

It even tracks what you type into the search bar—even if you don't hit enter.

Every moment you spend on the app is a data point, which is fed into Meta's larger network to refine the ads you see, the posts you're recommended, and even the people you're encouraged to follow.

Facebook

Facebook is the godfather of surveillance capitalism. It's built it's entire empire on data collection.

Beyond in-app behaviour, Facebook tracks:

- Your browsing history even off the platform
- Your app usage via third-party integrations
- Your location down to geofenced behaviour patterns
- And conversations you have near your device (microphone permissions have long been suspected of passive listening, though Facebook denies this.)

Its infamous "shadow profiles" even track non-users, collecting information about people who have never signed up for Facebook at all.

TikTok

TikTok's data collection practices have raised global concerns—including investigations by governments and cybersecurity experts.

The app has been shown to collect:

- Keystroke patterns
- Clipboard contents
- Precise location data
- Device identifiers
- Voice and facial biometrics

What makes TikTok different is the speed and precision of its algorithm. It doesn't take months to learn you—it takes hours. By analysing how long you watch a video, how quickly you swipe, and even the facial expressions you make while watching, it begins shaping your feed with stunning accuracy.

YouTube

Whilst not considered a form of social media, YouTube also tracks your entire watch history, including:

- What you click
- What you pause
- What you rewind
- What you stop watching after a few seconds

It uses this to build content funnels that guide your belief systems, particularly through recommended content. You start watching guitar tutorials. A few clicks later, you're down a political rabbit hole or conspiracy theory chain—not because you searched for it, but because the system chose the next step for you.

Each of these platforms operates under the same principle. Track as much as possible, learn what triggers you, and feed you more of it. The goal isn't to inform or entertain. The goal is to predict and control.

You're not just being watched. You're being studied.

Your Data is the Product

Every move you make on social media is tracked. What you click, how long you pause on a video, who you DM, which emoji you react with, what time you're most active, what you skip, rewatch, or ignore. Every one of those actions is logged, interpreted, and used to build a behavioural profile—an ever-evolving version of you, designed to sell you things and shape your feed. Even what you don't do is data.

- Didn't click that post? They noticed.
- Scrolled past that ad? It's logged.
- Typed something in the search bar and deleted it? Still recorded.

Nothing is wasted, because everything is valuable. Remember, if you're not paying for the product, the likelihood is, you are the product.

The platform's job isn't to connect you with friends. It's to learn everything about you—so it can predict what you'll do next, and sell that prediction to advertisers, brands, political campaigns, and data brokers. Your behaviour becomes a profile. That profile becomes a target, and that target becomes a monetised object, bought and sold in the background while you think you're just "checking your feed."

You are not the customer. You are the inventory.

Location Tracking: Your Digital Shadow

You don't have to tell your phone where you are. It already knows. In fact, it knows where you were yesterday, what time you arrived, how long you stayed, and how often you go there. That's not science fiction. That's standard operating procedure.

When you download an app—any app—you've likely seen the permission request: "Allow location access?" Most people tap "Yes" without a second thought.

After all, what's the harm?

But here's what that "yes" really means:

- Your exact GPS coordinates are recorded—often down to within a few meters.
- Your movement patterns are tracked over time.
- Your phone can determine when you're walking, driving, or stationary.
- It can identify frequent locations—your home, your workplace, your child's school.
- It can cross-reference that with time stamps, painting a complete picture of your daily routine.

Now imagine that data being sold. To advertisers, to insurance companies, to political campaign firms, to surveillance vendors, to data brokers who combine it with your online activity, purchases, and search history to build an incredibly detailed profile of who you are—not just where you go.

And here's the part that surprises most people: Even if you don't allow location tracking, your location can often still be inferred—through things like:

- Nearby Wi-Fi networks
- Bluetooth beacons
- IP address triangulation
- Your proximity to other people's phones that do allow location access

In short, you don't have to opt in to be dragged in. You are constantly emitting a digital shadow—a breadcrumb trail that never fades, never stops growing, and never really belongs to you. Once a system knows where you go, it starts to predict where you'll go next.

Do you remember that first time you got into your car to go somewhere, and Apple Maps predicted where you were going? Do you recall the feeling the first time that happened?

Once it knows your vulnerabilities—where you spend late nights, how often you visit a pharmacy, when you drop the kids off at school, when you're coming home from a work trip—it can exploit them. All of this… Just because you wanted an app to share photos with friends and family…

That's the cost of access.

The Privacy Illusion

Many platforms love to talk about privacy. They give you toggles and checkboxes. They show you controls. They throw around phrases like "you're in charge." But what they rarely say is this: Most of the data has already been collected. Much of it is inferred—based on what you do, who you follow, where you go, and how others like you behave.

Your data doesn't just describe you. It defines you—in the eyes of the algorithm— and once your identity becomes a prediction model, you no longer see the internet as it is. You see what the system wants you to see.

Who Else Is Watching?

You probably assume that your data stays with the app you gave it to. It doesn't.

Behind every 'Accept All' click is an entire data supply chain—an invisible economy where your information is bought, sold, repackaged, and sold again.

Sometimes hundreds of times.

You are not a single file. You are a mosaic—stitched together by companies you've never heard of, operating in offices you'll never visit.

These companies don't need your name to know who you are, because a name is just an identity. They track your patterns, and a pattern is behaviour. Behaviour is predictive—which makes it profitable.

That's what the data brokers are after. They build detailed behavioural profiles— then sell them to advertisers, political campaigns, health insurers, retail giants, and even government agencies.

- They know what you buy
- What you believe
- Who you live with
- Whether you're likely to vote
- Whether you're in debt

- Whether you're struggling with addiction
- Whether you're pregnant—often before you've told anyone else.

Once your data enters this shadow economy, you don't own it anymore. There's no "delete" button. There's no customer service line—and for young people—the consequences are even more dangerous.

Teenagers are joining platforms at younger and younger ages, often without any meaningful guidance or supervision from parents or adults.

- They post impulsively
- They imitate trends
- They upload things they later regret.

Not out of malice—but because they're still learning how to be human.

However, the internet doesn't forget. Once a photo is shared, a video posted, a comment written—it can be copied, downloaded, spread, and it's often out of their control from that point on.

Young people are now growing up in a world where every mistake can become permanent. Where a bad decision at 13 can be screenshotted, shared, and resurfaced at 23. Where consent is disregarded. Where one post—shared by a "friend"— can lead to a spiral of bullying, shaming, and a mental health collapse.

And it's happening all the time:

- Kids being filmed without consent and uploaded for laughs
- Fake profiles created to harass and humiliate
- Private moments turned into public content
- Cruel comments piling up while parents are unaware it's even happening.

This isn't just a safety concern. It's a human rights concern.

Adolescence is supposed to be a time of growth, of making mistakes, of learning who you are—without the fear of public exposure and permanent judgment. Instead, we've created a culture where kids are afraid to exist outside the algorithm's gaze, and where digital humiliation has become a modern form of violence.

No child should be shaped by shame. No teen should be stalked by a mistake they made before they understood the consequences. No young person should be afraid to grow up because the internet might never let them move on.

Yet, for millions of them...

This is the reality.

Brian Is Not Okay
(A Disturbing Reflection)

I hadn't seen Brian in months. He'd gone quiet. Unusually quiet. So I dropped by.

When I walked into his house, it was dark. No lights. No sound. Just stillness and a strange kind of static tension in the air.

The place was a mess—dishes everywhere, curtains closed, piles of clothes in corners like they hadn't moved in weeks.

"Brian?" I called out... Nothing.

Then I heard it.

"Ssshhh…"

It came from deeper in the house. I followed the sound and found him, crouched in the corner of the living room, back against the wall, head buried between his knees. Rocking. Slowly. Over and over.

"Brian," I said quietly, "what's going on?"

He looked up at me—and the fear in his eyes was something I'll never forget.

"They're listening," he whispered. "All the time."

"Who's listening?" I asked.

"They are," he said. "The phone. The TV. My speakers. The apps. They know what I'm doing. They hear what I say. I think something and I see an ad for it. I mention something to someone and it shows up on my feed."

He started tapping his head with the palm of his hand.

"I don't even say it out loud sometimes. I think it—and it finds me. I can't get away. I can't stop it. They're in my walls, my roof. They're in my head!."

"I tried to calm him down.

"Bro, you're spiraling. You just need to sleep. Maybe unplug for a while."

He grabbed my wrist.

"Unplug from what? Life? Society? It's everywhere. It's normal now. They're not even hiding it."

He started rambling about algorithms, microphones, camera access, keystroke tracking, how everything he does—online and offline—is being collected and used against him.

"They know when I wake up," he said. "They know when I'm alone. They know when I'm sad. They know what kind of person I am—and they use it. They use me. I'm not a person anymore. I'm a user. A target. A test subject."

He wasn't yelling. He wasn't panicked. He was resigned. He had accepted defeat.

Brian Is Not Okay (How Times Change)

If you heard a story like this twenty years ago, you'd assume Brian was suffering from paranoid delusions. He would have been diagnosed with a mental health condition. Institutionalised. Treated.

But now? Now Brian is describing the default reality for an entire generation. Everything he's saying is true, and worse—it doesn't even sound crazy anymore, because we've been conditioned to accept surveillance. To expect it. To joke about it. To shrug at it. To let our children grow up inside it.

The generation born after 2010 doesn't remember a world without this. Without tracking. Without targeted ads. Without their faces scanned and categorised before they've even learned to speak.

Once upon a time, it was only criminals who were fingerprinted—processed after committing a crime, catalogued by law enforcement, and tracked for public safety. Now, we are all fingerprinted. Not because we've done anything wrong, but because we carry a smartphone.

The fingerprint is just the beginning. You are now identifiable by:

- Your face (facial recognition)
- Your voice (vocal fingerprinting)
- Your typing rhythm (keystroke dynamics)
- Your walking pattern (gait recognition)
- Your device ID, and this is just the tip...

Sadly, what used to be a red flag for mental illness has become a feature of modern life—and if that doesn't disturb us—if we've stopped questioning it—then maybe we should all be a little more like Brian.

Opting Out Isn't Enough

You can change your settings. You can clear your cookies. You can log out. But the truth is, surveillance has become the foundation of the internet as we know it.

It's no longer a bug in the system. It is the system. And if we want to reclaim our freedom, we have to stop pretending this is just about convenience or personalisation. It's about control.

The first step toward taking it back is knowing: You are being watched, because your attention is valuable, as is your data.

Chapter 4:
Terms and Conditions – The Legal Trap You Didn't Read

It's the same dance every time. A new app. A new device. A new platform.

A wall of text appears—long, dense, and unreadable. At the bottom is a checkbox and a shiny button that says "I Accept."

You click it. Of course you do—because if you don't, you can't use the app. You can't connect. You can't participate. No checkbox, no access.

And just like that, you've entered a legally binding agreement—one that most people never read and even fewer understand.

This is not convenience. This is conditioning.

The Consent Illusion

We like to think of ourselves as autonomous, rational adults—fully capable of making informed decisions about our lives. But the digital world has reduced this autonomy to a single act: clicking a box.

That checkbox is meant to symbolise consent. But it's not consent in any meaningful way. It's coercion disguised as choice. You either agree, or you get locked out of the modern world. There is no negotiation, no discussion, no ability to pick and choose which clauses apply to you. It's all or nothing.

And the companies know you won't read it.

A Legal Shield Built From Apathy

Terms and conditions documents are legal armour for tech companies. They're not written for your benefit. They're written to protect the platform from liability while giving it the broadest possible rights to use, store, share, sell, and modify your data.

These documents are often:

- Dozens of pages long
- Written in legalese
- Designed to obscure rather than clarify

And yet, by clicking "I Accept," you've agreed to:

- Letting your data be collected, stored, and shared
- Granting perpetual licenses for your content
- Allowing AI systems to analyse your behaviour
- Accepting binding arbitration that limits your ability to sue

All without reading a single word.

The Cost of Clicking Without Thinking

We've become flippant with this ritual. We treat legal agreements like light switches—either on or off. But that habit is dangerous.

Because every "I Accept" you click reinforces a system that:

- Assumes your consent, even in ignorance
- Trains you to submit before you understand
- Normalises giving away rights in exchange for access

This behaviour may seem harmless now, but the implications for the future are profound.

We are raising a generation of people who believe that clicking "I Accept" is just part of daily life. Who think that giving up privacy, identity, and autonomy is just the way it is.

When that becomes normal, what else are we willing to give away?

When Terms Are Unreasonable—Except for Big Tech

In many industries, contracts that deceive, overwhelm, or confuse consumers have been legally challenged and voided by courts.

For instance, in 2024, the Australian Competition and Consumer Commission (ACCC) initiated legal action against a household name telecommunications company, for allegedly engaging in unconscionable conduct.

The telco was accused of signing up hundreds of vulnerable individuals, including a woman with cerebral palsy and ADHD, to multiple phone contracts they could not afford. In some cases, customers were allegedly entered into contracts over the phone without fully understanding the terms, leading to significant financial hardship.

Such practices have prompted regulatory bodies to enforce stricter consumer protection laws. The ACCC has emphasized that businesses cannot include unfair terms in standard form contracts, and any such terms found to be unfair by a court are void.

However, when it comes to Big Tech companies, the standard seems to disappear. These platforms routinely present users with lengthy, complex terms and conditions, often accepted with a single click, without real understanding or negotiation.

Despite the significant reach and risk of these terms, Big Tech has largely avoided the same level of scrutiny applied to other industries.

It's not consent; it's coercion through convenience, and despite the enormous reach and risk of these terms, they've seemed to become immune to the same legal scrutiny applied to smaller industries.

We've normalised something that would be unacceptable in almost any other business contract—and we've done it under the illusion that clicking "I Accept" means we actually understand what we've agreed to.

But the law has already shown, time and time again: When consent is uninformed, rushed, or strategically confusing—it's not valid consent.

So why do we keep treating it like it is?

The Future Is Being Written in Legal Fine Print

The deeper concern isn't just that terms and conditions are unread. It's that we are being conditioned not to care.

This is the real danger. Not what's buried in the document, but what's being buried in us: A mindset that expects no control, no transparency, and no say in the systems that govern our digital lives. That mindset doesn't just affect how we use apps. It affects how we see the world. We begin to believe that complexity excuses exploitation. That we are too small, too powerless to challenge anything. That it's easier to click "Accept" than to ask, "Why?"

Once we give up asking that question, we've already lost more than we know.

A Note to Parents, Teachers, and Caregivers

If you're someone responsible for shaping the mindset of the next generation—whether as a parent, teacher, mentor, or coach—there's something urgent to consider.

We teach kids not to talk to strangers. We teach them to look both ways before crossing the road. We warn them about hanging around with the wrong crowd.

Yet, when it comes to social media, we've largely gone silent. We hand over what are essentially super computers, and say, "Be careful," while the apps ask them to sign legal documents before they can even use the platform.

And they do. Without reading. Without pausing. Without understanding.

Would you advise a teenager to:

- Sign up as an organ donor without knowing they'd just given lifetime consent to any hospital in the country?
- Take out a $10,000 loan at 34% interest without reading the fine print?

- Agree to a medical procedure by signing a waiver that removes their legal rights if something goes wrong?

Of course not. And yet we stand by while young people click "I Accept" on contracts that grant sweeping access to their location, behaviour, voice, image, content, and identity—often with virtually no meaningful safeguards or oversight.

If they don't know what they're agreeing to, they're not giving consent. They're being trained to surrender their rights in exchange for convenience.

So if we're not helping them see that—who will?

The Illusion of Consent: When Minors Click "Accept All"

In most countries around the world, children and teenagers cannot legally enter into binding contracts. They can't lease a car. They can't sign a mortgage. They can't get a credit card.

Why? Because the law recognises that minors lack the legal capacity and maturity to fully understand the implications of such agreements. And yet—every day, millions of children and teens are silently entering binding digital contracts. Not with lawyers, not with advisors, but with tech giants.

Through nothing more than a button that says: "Accept All."

If a minor clicks "Accept All," and they're not of legal age to provide informed consent, then the company collecting their data is not legally entitled to it. At least, not under traditional contract law.

In most jurisdictions, contracts signed by minors are voidable—meaning the child can back out at any time, and the contract is not legally enforceable.

Yet social media platforms continue to:

- Accept these agreements at face value
- Treat minors as if they are legally consenting adults
- Harvest data just the same

Why? Because no one is stopping them.

The Loopholes They Rely On

a) Self-reported age: Most platforms "verify" age with a simple dropdown. If a 12-year-old says they're 16, the system accepts it.

b) Buried disclaimers: Platforms shift the responsibility to parents—often hidden deep in their legal pages.

c) "Service agreements" vs. contracts: Companies argue these aren't traditional contracts, but terms of use—further blurring the legal line.

But no matter how it's framed, the reality is this: A child is being asked to agree to a lifelong data relationship with a corporation—before they even understand what privacy means.

The Ethical Problem No One Wants to Face

The question isn't: Can Big Tech do this? It's, why are they allowed to?

We know these users are underage. We know their consent isn't informed. We know the platforms are collecting data that could follow them for life. And still, we do nothing.

A Call to Wake Up

If your child signed a credit card agreement without your knowledge, you'd challenge it. You'd cancel it. You'd call it what it is: invalid.

But the same child can download a dozen apps, click "Accept All," and hand over:

- Their face
- Their voice
- Their location

- Their conversations
- Their browsing history
- Their habits, emotions, and vulnerabilities

And the law shrugs.

This is not just a privacy concern. It's a systemic failure of protection, and until we face it, children will continue to be silently signed into lifelong surveillance contracts they never had the legal or emotional maturity to understand.

The big question is: Why aren't regulators holding platforms accountable for knowingly accepting invalid consent?

The Box You Should Have Never Checked

Every check box is a contract. Every contract is a surrender of something, and most of us don't even know what we've given away. The next time you see that checkbox, pause. Ask yourself:

What are they getting in exchange for what I think I'm getting? What am I giving away for access?

Because the more we normalise blind consent, the harder it will be to reclaim our freedom later. Especially when the next generation no longer remembers what freedom felt like in the first place.

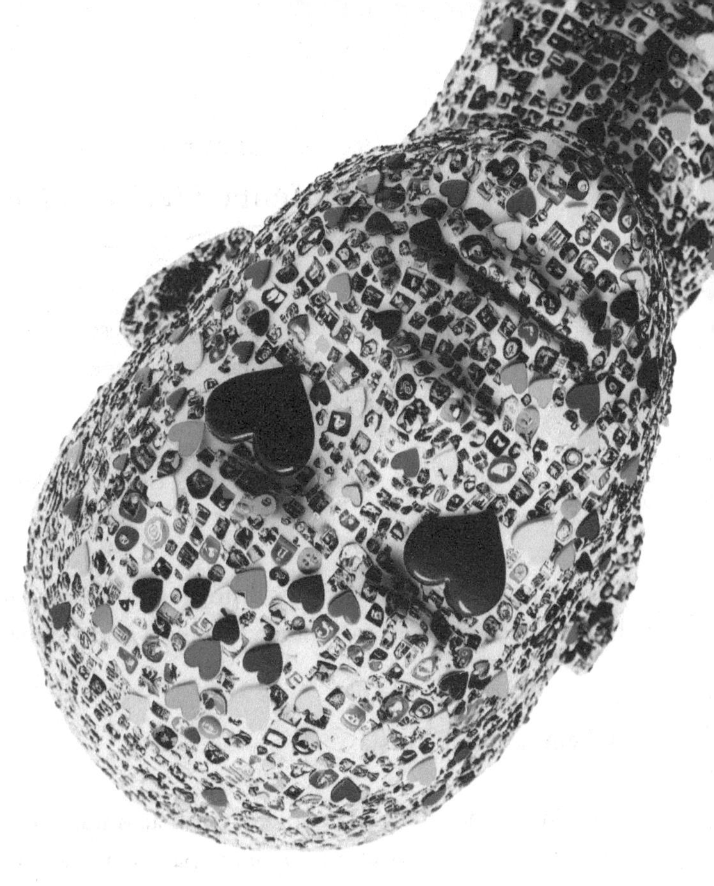

"WE WERE NEVER MEANT TO HEAR EVERY VOICE, NOR MEASURE OURSELVES AGAINST THEM."

ALT.LANE

Chapter 5:
The Mental Health Crisis

We are not okay. Look around. Anxiety, depression, insomnia, eating disorders, self-harm, suicidal ideation—mental health issues are rising across every age group, but nowhere more sharply than in young people.

While no single cause explains this surge, one factor keeps appearing at the center: social media.

It's not just that these platforms distract us. It's that they've fundamentally rewired how we see ourselves, each other, and the world around us.

A Generation in Crisis

Teen mental health has deteriorated at an alarming rate—particularly in the years since social media became a dominant part of adolescence. Kids are reporting more loneliness in a hyper-connected world. More self-hate in a time of filters and face-tuning. More stress, more confusion, and fewer tools to navigate a world that won't let them unplug.

But it's not just the mental toll. It's what they're being shown, sold, and told—every single day.

Social media has become a breeding ground for toxic comparison, impossible standards, and algorithm-driven manipulation, particularly in young girls. Children as young as 8 or 9 years old are being fed a steady stream of content—often from influencers, not experts—that tells them their skin isn't good enough. That their pores are a problem. That they need serums, gels, and anti-aging solutions meant for adult women.

And they're buying in—literally. The tween and teen skincare market is projected to surpass 200 billion by 2030.

What an absolute disgrace!

This isn't an accident. It's a marketing strategy, executed through TikTok trends, influencer collaborations, and "Get Ready With Me" routines that blur the line between play and performance.

Big beauty brands are using unqualified influencers to infiltrate the self-worth of young girls, subtly suggesting that their natural face isn't enough. That youth should already be working to preserve youth. That beauty requires effort, and effort requires product. So, whilst the big beauty brands have "A LOT" to answer for, social media willingly provides them with the platform to reach this demographic of consumer.

What was once a category reserved for mature adults—anti-aging creams, retinol, chemical exfoliants—is now lining the bathroom shelves of children.

Dermatologists are already raising alarms: This is not safe. This is not normal. This is not okay.

It's not just the skin irritation or allergic reactions. It's the message beneath it all: You are not enough, but you can buy your way closer.

We have to ask ourselves: When did childhood become a consumer category?

Why are we letting corporations sell insecurity to 10-year-olds? And what happens to a generation that learns they need to "fix" themselves before they even finish growing?

This is no longer about self-expression or experimentation. It's about a systemic erosion of identity and confidence at the hands of corporations that care more about customer lifetime value than psychological well-being.

A generation in crisis doesn't just happen. It's built. One post, one product, one manipulated moment at a time. Unless we start pushing back, the crisis will only deepen.

The New Disorder: Algorithmic Insecurity

Social media doesn't just trigger comparison. It manufactures it. Platforms constantly feed users content designed to keep them engaged—and nothing engages quite like inadequacy.

- Fitness influencers flaunting bodies forged by surgeries, enhancements, professional lighting and filters—while claiming it's all "discipline."
- Fake entrepreneurs selling get-rich-quick fantasies through "just follow my method" videos that are rarely based in reality, often peddled by people with no qualifications beyond charisma.
- Travel couples posing in Bali with captions like "just manifest it," while behind the scenes, the photos were staged, the moment miserable, the relationship fragile.

Then there are the darker corners of the feed—the ones that glorify dysfunction:

- OnlyFans creators counting stacks of cash, pitching sex work as a fast-track to wealth and independence.
- TikToks glamorizing sugar daddy culture.
- Influencers equating empowerment with selling their bodies for views and validation, under the guise of "self-expression."

These aren't just isolated cases. They're trends, and the algorithm promotes them because controversy and intensity drive engagement. The more outrageous, the more rewarded. The more damaging, the more viral.

And our kids are watching.

No Time to Be Human

Social media doesn't leave space for silence. For stillness. For boredom. For breath. We weren't meant to live like this. To be constantly interrupted. To have every quiet moment filled. To never truly arrive in our own lives because there's always something else to scroll, tap, or respond to. Yet, here we are—half-human, half-distracted. Living on autopilot. Trapped in timelines. Present in body, absent in mind.

Look around. On the train, at the bus stop, in waiting rooms, in restaurants, at traffic lights, in bed, even in bathrooms. Heads down. Eyes glazed. Fingers scrolling.

It's not science fiction. It's daily life. A society walking around like zombies, hypnotised by blue light and endless content.

If that feels dramatic—try this:

A Simple Test

Stop and observe your own household for one night.

- What time does everyone pick up their device?
- How often do people speak without making eye contact?
- How many screens are active at once?
- And most importantly: When was the last time everyone in your home was together, in the same room, doing the same thing—with no devices present?

Chances are, it's rare.

We've come to accept the sight of an entire family in one room, each person lost in a different screen, as normal. But it's not normal. It's a warning—because if we've lost connection inside the home—our most intimate space—how can we expect to reclaim it in the world outside?

Change doesn't start with tech companies. It starts with us. In kitchens, on couches, at dinner tables, in the quiet spaces where real conversation used to live.

We are slowly giving up our right to be human—not in one dramatic moment, but through a thousand unconscious scrolls. Until we recognise that it's happening in our homes, we'll keep pointing fingers outward... and never realise we're building the same problem behind our own doors.

The Unseen Consequences

What happens to a child who has been shaped by filters and algorithms more than by parents and teachers? What happens to a teenager who's learned that attention is currency and that silence is failure?

What happens to a human being who believes they are never allowed to just be—unless it's shareable?

You get what we have now:

- Panic attacks at thirteen
- Burnout at fifteen
- Body dysmorphia before puberty
- Social withdrawal in a hyper-connected world

We built the machine. Now we're watching it grind down the very people we thought it would connect.

This Isn't Okay

Let's stop pretending this is just how life is now. It's not normal for a teenage girl to compare herself to someone with a team of stylists, editors, and AI-driven beauty filters.

It's not okay that young boys are taught their value lies in wealth, dominance, or going viral.

It's not acceptable that children are absorbing sexualised, staged, false realities as their baseline for adulthood.

This is not okay! And the longer we accept this as the price of being "connected," the more we risk raising a generation that has never known what it feels like to simply be enough.

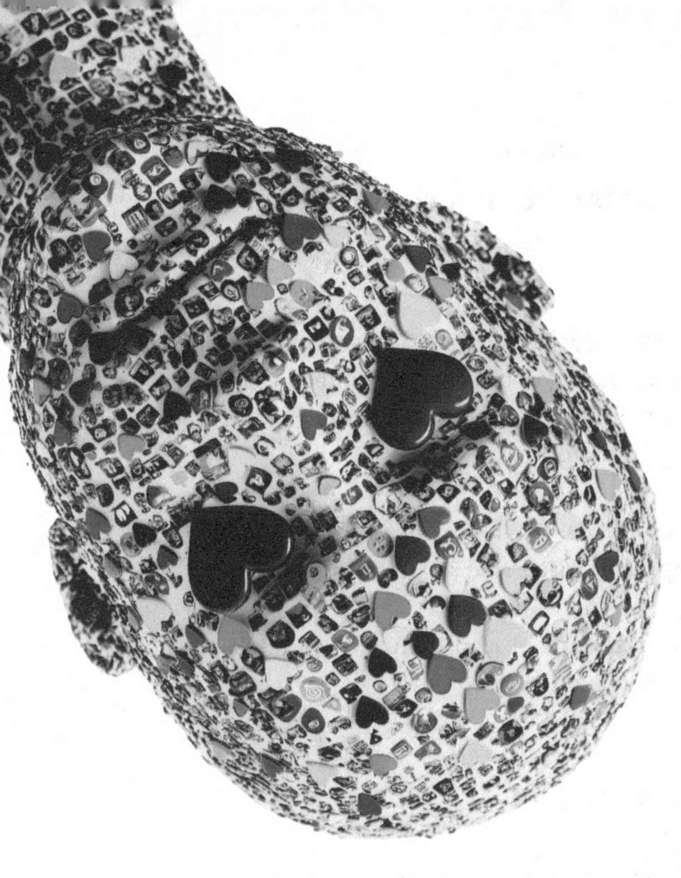

A GENERATION IN CRISIS DOESN'T JUST HAPPEN. IT'S BUILT. ONE POST, ONE PRODUCT, ONE MANIPULATED MOMENT AT A TIME.

Chapter 6:
Raised by the Algorithm

There used to be a time when children were raised by communities, parents, grandparents, neighbours, teachers, coaches. They inherited beliefs from conversations at the dinner table, learned boundaries through human feedback, and formed identity through real-world experiences.

Now? They're being raised by feeds. By "For You" pages. By trending soundtracks, 7-second loops, and AI-curated suggestions.

Children today are being shaped more by algorithms than by adults, and the consequences are only just beginning to show.

When the Feed Becomes the Parent

From the moment a child picks up a device, they're introduced to a world of nonstop influence—not from trusted role models, but from whatever content is most likely to go viral. And they're spending a staggering amount of time there.

Recent studies reported by Common Sense Media show that teens and young adults now spend over 90% of their free time in front of a screen—most of it on platforms powered by recommendation algorithms.

If you're a parent, and this is the daily reality in your home: Do you know what your son or daughter is consuming? Do you know what voices are shaping their views?

If you don't… but you let it happen anyway, then you are allowing the feed to take your place—and based on what we know is being promoted—sexualisation, outrage, superficiality, false wealth, and distorted realities—this is not acceptable.

So why are we letting it happen? Why are we handing over the role of mentor, guide, and teacher to a system whose only goal is engagement?

We wouldn't let a stranger walk into our house every night and whisper into our child's ear.

But that's exactly what the algorithm does, and we're doing nothing to stop it.

The Disappearing Role of the Adult

Ask any teacher, any parent, any psychologist—they'll tell you: kids are different now. Not just in their behaviour, but in their attention spans.

When everything comes in flashes, clips, swipes, and slides—nothing holds their focus unless it's extreme.

They've been trained by the scroll:

- Skip what's boring
- Pause only on what hooks
- Watch what shocks
- Like what triggers
- Move on

And what they pause on—the thing that makes them stop—is not just what the algorithm remembers, but prioritises. So the system feeds them more of what disturbed them. More of what confused them. More of what left a mark.

When every decision on social platforms is driven by attention data, the most extreme content always rises. The calm, the reflective, the educational—it gets buried.

This is why the adult voice is disappearing, because wisdom takes time—and time is something the algorithm doesn't grant us more of.

It does the opposite. It sucks it away.

What's Being Rewarded

Let's be clear: children learn by imitation, and right now, they're imitating influencers who:

- Stage fake drama to boost views
- Pose in luxury they don't own
- Promote sexual content for attention
- Act outrageously because it's rewarded

Young boys are being taught that fame comes from clout, controversy, and dominance. Young girls are being shown that worth comes from visibility, desirability, and sexual appeal.

None of this is accidental, and it's certainly not the natural evolution of our society. It's because once the algorithm detects what performs, it amplifies it.

And what gets amplified becomes the model.

Identity by Algorithm

Before the age of 18, a young person is still developing:

- Their sense of self
- Their values
- Their worldview
- Their emotional resilience

Now imagine that development being constantly interrupted by:

- Beauty filters
- AI-augmented voices
- Clickbait content
- Hate-filled comment sections
- "Authenticity" manufactured for clicks.

The line between real and artificial becomes blurred. Children no longer know what's genuine—because everything is content. And if everything is content, then they become content too. Their emotions. Their relationships. Their trauma. Their joy.

All of it packaged for public consumption.

The Quiet Collapse of Childhood

There was a time—not long ago—when childhood meant freedom.

Freedom to be bored. To play outside. To imagine wildly. To make mistakes without an audience. To grow up slowly. That time is slipping away, and the collapse isn't loud, it's quiet. Invisible even.

It doesn't look like a crisis. It looks like a kid on a couch, eyes glued to a screen. It looks like a toddler swiping before they can speak. It looks like a family dinner where no one speaks at all.

But make no mistake: something sacred is being lost.

Social media didn't just speed up time. It compressed childhood into a hyper-curated highlight reel—where the expectation is to look older, act bolder, and gain followers faster.

Ten-year-olds now:

- Know how to edit photos and apply filters
- Imitate influencer dances
- Speak in hashtags, soundbites, and viral phrases
- Worry about their appearance before they've even hit puberty

Instead of building treehouses, they're acting like they are building personal brands. They're skipping the parts of life where you stumble privately and grow quietly.

The end of innocence is happening one scroll at a time, with no gatekeeper, no soft landing and no emotional filter.

A child can be exposed to:

- Violence
- Pornography
- Racism

- Self-harm
- Gendered hate
- Disinformation
- Suicidal ideation
- Extreme political content

...all before they're old enough to drive.

And the content doesn't ask, ***"Are you ready to see this?"*** It just shows up. That's the world we've handed to our kids, and we call it "normal."

Is this what childhood was meant to evolve into? Unlikely.. Childhood is meant to be a time of imagination, not monetisation. Of discovery, not documentation. Of connection, not comparison. But we handed kids a portal to the adult world, and now we're watching the collapse in real-time.

So what are we going to do about it?

This isn't just a parenting problem. It's a societal failing, because the systems designed to protect children—lawmakers, educators, even families—have been too slow, too distracted, or too complicit to act.

While we debate policy and screen time limits, another generation is quietly losing their only chance to be young, free, and unburdened.

This chapter doesn't end with a solution. It ends with a question: How many childhoods are we willing to sacrifice before we change the system that's stealing them?

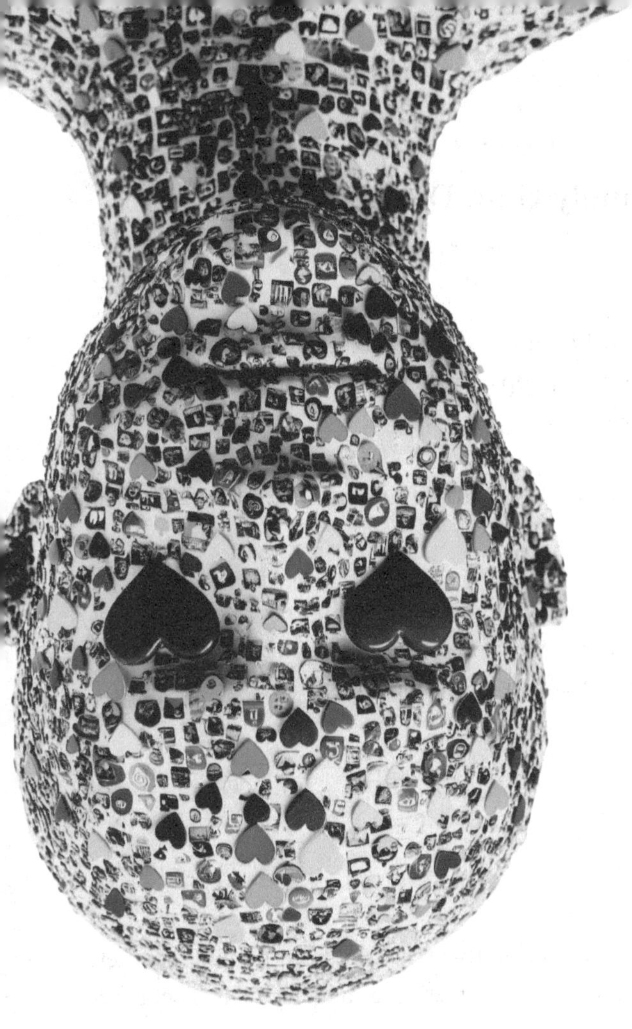

"WE ALL SIT TOGETHER, BUT SEEM TO BE WORLDS APART."

ALT.LANE

Chapter 7:
The Family Unit, Disrupted

There was a time when the family home was a sanctuary. A place of safety, privacy, and presence. A place where people could eat together, talk together, argue, laugh, cry, and grow—together.

Today, that sacred space is under siege. Not from outside threats, but from inside screens.

We've allowed devices into every room. We've placed them in the hands of every family member, and slowly—quietly—we've allowed them to become a fifth presence at the dinner table. Uninvited. Unquestioned. Constant.

And it's changing us.

Alone, Together

Walk into most homes today and you'll find this scenario: Four people together, but worlds apart. A parent scrolling emails. A teen watching TikToks. A child deep in YouTube. Another parent doom-scrolling the news.

The living room is no longer a shared space. It's a digital waiting room where everyone is physically present but emotionally elsewhere—and the apps love this. Because the more isolated we are, the more dependent we become. The more divided our time and attention, the easier we are to manipulate. So we retreat further into our devices, even while longing for connection.

But it goes deeper than silence. Social media hasn't just disrupted our conversations—it's replaced their subject matter. A son used to come home and talk about what happened at school. Now, he walks in and says, "Dad, look at this video." Instead of sharing something personal, he shares something viral.

A daughter used to bring up what happened in class, or how she felt that day. Now, she shows a TikTok trend that got 5 million likes that she is going to try to recreate. And that's if they are even talking to their parents at all…

Conversations that once revolved around each other's lives are now built around the lives of strangers online. The family has become an audience—watching, reacting, and scrolling through someone else's script. Real stories are being pushed aside in favour of second-hand distractions.

This shift, subtle as it is, marks the fading heartbeat of true connection.

When Family Becomes Background Noise

We've become so used to seeing our loved ones distracted, we've stopped expecting them to be engaged. And it starts early.

Before many children even learn to speak, they've already learned to swipe. Screens are placed in front of toddlers to keep them quiet, entertained, or calm—and in doing so, we unintentionally disconnect them from the very world that helps them grow.

This is how isolation begins. Not with loneliness—but with silence. Not with absence—but with disconnection in proximity. The more isolated we become, the more we retreat further into our screens for comfort—reinforcing the very problem we're trying to escape.

Research has shown that the more time a person spends on screens, the more likely they are to experience social isolation with measurable effects such as:

- Increased anxiety and depression
- Reduced empathy
- Weakened emotional resilience
- A diminished sense of identity and belonging

Now multiply that across a generation raised in digital spaces instead of human ones. We're not just risking less family connection—we're raising a generation that might not even know what real connection feels like.

If your earliest memories involve quiet dinners, disconnected parents, and solitary screens, what do you grow up expecting from others? What does "normal" feel like?

The longer we accept this emotional drift as inevitable, the harder it becomes to close the distance. Slowly, the people we love will become background noise—visually present, emotionally muted, and mentally elsewhere.

The Hijacking of Household Dynamics

The influence isn't just passive. It's strategic. Many apps don't just listen to individuals—they map household behaviour:

- Who's connected to what device
- What rooms different users spend time in
- What voices are detected
- Which products are mentioned
- What relationships exist between users

By doing so, they infer household roles, routines, and power dynamics—then tailor content and ads accordingly.

It used to be that the family home was the one place free from outside eyes. Private. Personal. Sacred. Now, that space is being infiltrated. Not with brute force. Not with microphones hidden under floorboards. But with something much subtler—and somehow far more accepted.

We clicked "Accept All," and in doing so, we gave apps and devices permission to monitor us inside our most private spaces. Smart speakers listening for "wake words." Phones tracking which room you're in. Apps requesting microphone and camera access "just in case" — Yeah, sure!

So what are they doing with this access?

They're mapping household behaviour. They know who lives there, who speaks the most, who gets ignored. They can detect which voices are dominant, when you tend to argue, how many people are home, and which devices are used by whom.

This is surveillance. Not safety. Not service. Surveillance.

Imagine walking into your home and finding a tiny microphone hidden behind a photo frame. Or discovering a camera in the corner of your ceiling, silently recording your every move. Planted by someone else. Without your knowledge.

You'd be furious. You'd feel violated. You'd tear the house apart looking for more. We all know the feeling. Many of us have checked for hidden cameras in hotel rooms. Wondered if we were really alone, or if there is a camera behind the mirror in the bathroom, or a listening device in an air vent. We have thought twice before saying or doing something sensitive in a place that didn't feel truly private.

And yet… We install devices that do exactly that in our homes. Willingly!

Why? Because it comes with a weather forecast, or a sleep tracker, or a recipe suggestion, or it plays music, or controls the lights.

All of a sudden, constant surveillance feels like a feature.

In 1993, the Hollywood movie Sliver showed us a stalker who secretly monitored every tenant in a luxury apartment building. He watched their conversations. Their habits. Their intimacy. It was disturbing, illegal, and universally understood to be morally reprehensible.

30 years later, that same setup exists in millions of homes—only it's been rebranded as convenience. It doesn't feel like surveillance anymore. It's branded as "smart living," because we've been conditioned to trade privacy for personalisation.

We've allowed the deepest kind of intrusion into our personal spaces—Not because we had no choice, but because we were too distracted to question it.

Now, the same kind of surveillance technology that was once used to send people to prison…is accepted in our homes as a service.

The Influence You Didn't Authorise

Families don't just want proximity. They want connection. They want:

- Safety—to know that home is a place where you are protected, respected, and truly seen.
- Presence—not just physical, but emotional presence; knowing someone is really there with you, not lost in another world behind a screen.
- Engagement—genuine conversations, shared experiences, memories built together, not separately.
- Pride and closeness—the joy of watching each other grow, stumble, learn, and succeed—not through a filtered highlight reel, but through real, messy, human life.

This is the fundamental joy of creating a family unit: To be close. To be proud of one another. To be woven into each other's lives, not simply coexisting under one roof. But that's not where we are at.. A new influence has entered the home—one that no parent voted for, no sibling welcomed, no child fully understood.

It speaks louder than we do, suggests what to wear, how to speak, what to care about. It dictates attention, rewires values, and rewards conformity to trends rather than individuality. It replaces real encouragement with performative validation. It subtly, day by day, reprograms the people we love.

Now, instead of sharing real dreams and struggles across the dinner table, we trade viral videos, memes, and half-heard stories interrupted by notification pings. Instead of kids absorbing the values of their parents, they're absorbing the curated scripts of strangers—influencers who are paid to sell lifestyles, not live them.

Home used to be the place where you learned who you are. Now, for many, it's just the backdrop to another scroll session, and the cruel irony is: We haven't lost the desire for connection. We've simply been distracted from fulfilling it.

The need is still there. The longing is still there. But unless we start fighting to reclaim it, it will continue to slip.

So now, imagine a home where dinner is filled with conversation instead of scrolling. Where evenings end with laughter echoing down hallways, not the glow of separate screens behind closed doors. Where family members truly see each other—unfiltered, un-curated, unhurried. Where validation comes from a proud look across the table, not a heart icon on a post. Where love is shown not just in words, but in full attention—in choosing to listen to each other.

This isn't a fantasy. It's still possible. But it takes awareness. It takes intention. It takes remembering that connection must be cultivated, not assumed. Because presence isn't just something you have. It's something you give. And there is no app, no trend, no influencer in the world that can replace it.

Chapter 8:
The Illusion of Freedom

We think we're in control. We choose the apps we download. We decide who we follow. We tap what we like. We believe we're navigating social media on our terms.

But the truth is far more uncomfortable:

You are not the one steering. You are being nudged. Guided. Filtered. Shaped. Not by a person. Not even by a team. But by a machine—one designed to make you think your choices are your own.

"I Can Stop Anytime"

We tell ourselves we're in control. That we can quit any time we want. That we choose to pick up the phone, to scroll, to post, to like. But choice isn't real if the compulsion has already taken root.

The truth is: Most people have no idea how deeply entangled they are—until they try to let go.

So here's a challenge: Pick one night this week. Declare it a "No Device Night." Not forever. Not as a punishment. Just an observation.

The rules are simple: Phones go away. Tablets go away. Laptops go away.

If you're with family, roommates, or friends, you can keep the TV on—but only if you agree on something to watch together. Then watch what happens.

Do you or the people around you: Get restless? Start reaching for pockets or glancing toward the place the phone should be? Show signs of irritation, frustration, or withdrawal? Get moody or short-tempered without fully understanding why?

Because these are the telltale signs of addiction: Restlessness, anxiety, mood swings, cravings for the dopamine hit of the next notification. It's not about enforcing discipline. It's about witnessing the invisible chains you didn't realise were wrapped around your mind.

If it feels uncomfortable—good. That discomfort is your first glimpse of freedom. Because you can't fix what you refuse to see.

The Algorithm Is Driving

Once upon a time, algorithms were simple. You clicked on a video about baking bread, and you were shown another video about baking bread. The system made suggestions based on obvious patterns—basic, surface-level connections.

However, that's not what's happening anymore.

Today's algorithms don't just react to what you like. They predict what will keep you there the longest.

They analyse your:

- Watch time
- Scroll speed
- Pauses
- Rewatches
- Comment habits
- Emotional responses based on facial expressions and text sentiment.

They don't show you what you want. They show you what you can't resist, and nowhere is the evolution of the attention economy more visible than on YouTube.

Once a simple platform for sharing homemade videos and niche content, YouTube has become a global entertainment machine—with the algorithm at the helm, and in this system, the loudest, fastest, most exaggerated content wins.

Just look at the gaming genre, one of YouTube's largest youth-driven categories.

Content Creators scream at the top of their lungs in reaction to relatively minor in-game events. "OH MY GOD!!!!, OH MY GOD!!!!" distorted audio, crackling through speakers.

Facial expressions are exaggerated. Voices are pitched. Editing cuts are rapid and relentless.

The content isn't just about gameplay anymore—it's a performance. Every second is choreographed for retention. Every reaction is dialled up to 100. It's not enough to play the game—you have to turn the volume up on your personality so high that viewers can't look away.

Other creators—especially in lifestyle, or pop culture niches—do the same. They splice their videos into ultra-short intervals, never letting a scene breathe. Every two to three seconds, there's a cut, a sound effect, a meme overlay, a flashing visual.

It's not because viewers demand it—it's because the algorithm rewards it.

This format wasn't born from artistic choice. It was born from a single need: Retain attention! And the more the algorithm favours this style, the more creators copy it. Slowly then, without realising it, we lose tolerance for anything that moves slower than a TikTok scroll.

Now, today's algorithms operate on a new principle: Predictive behavioral analytics. This involves using data analysis and behavioural design to strategically influence desired changes in behaviour.

They don't just predict what you'll click. They gently, invisibly, push you toward becoming the kind of person who will keep clicking. They expose you to repeated types of content. They feed you fear, outrage, aspiration, insecurity—whatever emotion prolongs your session.

What started as "recommended for you" has turned into "engineered for your entrapment."

Curated Beliefs, Manufactured Opinions

We like to think our opinions are our own. That we come to conclusions through logic, experience, and independent thought. But in the digital world, beliefs are curated just like content. You are shown certain videos. You are recommended certain posts. You are nudged toward certain conversations. Not because they reflect the truth—but because they increase your engagement. Once you click on a certain type of content—whether it's political commentary, conspiracy theories, or niche worldviews—the system feeds you more of it. Not just slightly more—exponentially more.

You clicked on one video questioning the government? Here's five more suggesting the government is actively plotting against you. You watched a video supporting your political side? Here's a dozen more showing why the other side is evil, stupid, or dangerous.

So, what started as a casual curiosity—becomes a radicalised worldview. Without you even realising it, your mental world shrinks. You stop seeing alternate views. You stop hearing reasonable disagreement. You stop questioning the narrative. Because everything around you—the videos, the comments, the "suggested for you" posts—reinforces the idea that you are right, they are wrong, and compromise is betrayal.

We are seeing conspiracy theories explode in a way no previous era has witnessed. Not because more people are suddenly paranoid—but because the machine rewards paranoia. It rewards fear, suspicion, anger, and distrust. It pushes people further into rabbit holes where every new piece of information is distorted to fit the pre-approved narrative.

Once fringe theories—"vaccine microchip tracking" or "flat earth" beliefs—have found new life and mass audiences. Not because they are convincing in themselves—but because once the right emotional button is pushed, confirmation bias takes over, and the algorithm ensures you see only what feeds that bias.

The same is happening in politics. Whether far right or far left, platforms encourage users to become more extreme, more hostile, and more unyielding. The more polarised you are, the more likely you are to stay online defending your side.

Extremes aren't a byproduct of the system. They are the goal, because moderation, nuance, and critical thinking don't pay—and now the world looks like it's gone crazy. To those standing outside the algorithmic pipelines, it feels like everyone has gone mad.

You look around and wonder:

- When did facts become optional?
- When did discussion become warfare?
- When did disagreement become moral failure?

The answer isn't a sudden collapse of intelligence or civility. The answer is conditioning.

When you live inside a system designed to reward outrage, division, and emotional manipulation—you get a society that reflects exactly that. So unless we recognise that our feeds aren't mirrors of the real world—but manufactured funnels of thought—the polarisation will only deepen. We are not living in the age of information. We are living in the age of manipulation, and the sooner we wake up to that fact, the sooner we can start thinking for ourselves again.

You are what you eat!

Choice Is Not Freedom

Just because you can choose what to post, doesn't mean you're free. Just because you can follow who you want, doesn't mean you're in control.

Freedom isn't defined by what you're allowed to do. It's defined by what you're able to resist, and social media is built to break that resistance.

The colors, the notifications, the buzzes, the loops—they are all designed to interrupt your focus, override your decision-making, and bring you back under control.

So ask yourself: If you can't stop using something that's making you anxious, if you can't scroll without comparing, if you can't step away without fear of missing out—Is that really freedom?

Social media sells itself as liberating. It promises you a voice. It gives you a platform. It says, "You are the creator now." But the truth is, you're only free to create within the boundaries of what the algorithm will promote.

If your content aligns with the metrics—great. If it doesn't, it vanishes, and so creators start shaping themselves to suit the feed:

- Louder
- Edgier
- More filtered
- More extreme

This isn't self-expression. It's self-alteration for digital approval. We're not creating freely. We're performing for machines.

You were never meant to be controlled—you were born with the ability to choose, to create, to reflect, to wonder, to connect – to think critically. But these platforms are being built to replace those human instincts with automation. To turn curiosity into compulsion. To turn autonomy into reaction.

And it's happened so subtly, we didn't even see it coming.

Chapter 9:
What You're Missing While Scrolling

We talk a lot about what social media does to us—the addiction, the comparison, the anxiety—but there's something just as important that's easier to overlook: what it displaces. Because every hour spent scrolling is an hour not doing something else. Something real. Something present. Something human.

The greatest cost of the attention economy isn't just what it takes from your mind. It's what it takes from your life.

The Lost Hours

Let's start with the obvious. Just search on google: "On average, how long are people spending on social media today?" According to numerous global research groups, the average person now spends 2 hours and 23 minutes a day on social media. That's roughly 17 hours a week, or 110 / 8hr work days per year—spent watching strangers live their lives while yours quietly waits in the background.

Multiply that over a decade, and you've handed over more than a year of your life to the scroll. And for what? Recycled memes and arguments in comment sections? Are you joking! This is time you'll never get back, and unlike money, you can't earn it
again.

The Death of Presence

Presence used to be a natural state. It wasn't something you had to think about—it just was. You were where you were, with the people around you, fully available to the moment. Now, presence is a rare commodity. A luxury. Something we chase through meditation apps and mindfulness podcasts—because we've lost the ability to just be where we are.

What makes this loss so devastating isn't the big moments we miss. It's the small ones. The micro moments.

- The glance from a child that says "did you see that?"
- The shared laugh that didn't need to be spoken.
- The subtle change in a loved one's tone that hinted they weren't okay.
- The moment the sun hit the window just right, and reminded you—however briefly—what it means to be alive.

So we need to remind ourselves that presence is not a renewable resource. Once a moment is missed, it's gone.

When you're constantly living somewhere else—in another tab, another feed, another world—you stop being part of your own. Your memories fade faster. Your experiences feel thinner. Your joy feels delayed, second hand, derivative.

We are slowly, collectively learning how to be half-present in everything, and the cost is staggering. Because presence is where connection happens. It's where we grow empathy. Where we hear what's not being said. Where we remember who we are—not through reflection, but through experience.

If we let those moments keep slipping away, we'll wake up one day surrounded by everything that matters—but unable to feel it.

Creativity Without Space

You were born to imagine. To create. To build. To reflect. But creativity doesn't emerge from chaos. It requires space, stillness, silence, boredom even. Yet boredom is the one thing social media will never allow. In fact, it has been engineered out of existence. The second boredom creeps in, there's a scroll to soothe it. A feed to refresh. A video to autoplay.

Boredom is not the enemy! It is the womb of invention. The quiet waiting room where ideas are born.

Throughout history, stillness was not a limitation. It was the precondition for genius.

- Albert Einstein documented numerous "thought experiments"—which were long stretches of time where he allowed his mind to wander freely with no distractions.

- Nikola Tesla developed complex inventions entirely in his imagination before ever building a prototype. He often sat in solitude, visualising machines in full detail, down to how they would move.
- J.K. Rowling first envisioned the world of Harry Potter during a delayed train ride. She had nothing to distract her—no phone, no screen—just her thoughts and a moment of stillness.
- Isaac Newton formulated the laws of motion and gravity while spending a year in isolation during the Great Plague. No distractions. No external noise. Just a quiet orchard and a falling apple.

So, what would have happened if they had been scrolling?

The Death of Boredom Is the Death of Depth

When we eliminate boredom, we eliminate the conditions that foster depth, curiosity, and originality. Social media fills every gap, and when there are no gaps, there is no room for new ideas to enter.

- You can't hear your own thoughts if you're constantly listening to someone else's.
- You can't build something truly new if you're always consuming recycled trends.
- You can't feel the tug of your imagination if your attention is being hijacked every 30 seconds.

We need the empty spaces back, we need to protect empty space, we need to reclaim the blank page. Because in those quiet, unremarkable moments—on a long walk, in a waiting room, staring out the window, sitting in stillness with nothing to do... that's where real creativity whispers, and that whisper can only be heard when the noise stops.

Relationships With Distance

We were meant to connect face-to-face. To hear tone. See eyes. Share energy. But social media has introduced a false proximity—the illusion of being close, without the depth.

- A like is not the same as a conversation
- A comment is not the same as presence
- A DM is not the same as showing up

Yet—for many of today's youth—this is what friendship has become. We now live in a world where "connection" is measured in notifications, not in eye contact. Where emojis replace emotion, and text threads replace time spent together. Where teens and tweens talk about having "best friends" they've never met in person. Friends they know only through usernames, filtered selfies, and curated content.

For many, this isn't unusual. It's the norm. But something essential is being lost. Digital connection is easy. It's instant. It's convenient. But it's also shallow.

You don't see someone's body language. You don't hear their tone. You don't sense when they're holding something back. You miss the pauses, the awkward silences, the unspoken truths that only real presence can uncover.

Real relationships are messy. They require time, patience, and effort. They involve showing up when it's inconvenient. Sitting in silence when someone is hurting.

Being present when nothing needs to be said at all.

Some young people are forming ongoing "friendships" with AI bots or influencers run by algorithms. These bots are programmed to engage, respond, flatter, and simulate care.

And it works.

To the human brain—especially a young, emotionally developing one, connection needs consistency—and replies feel like friendship, even when they come from a non-human source.

But that relationship doesn't know your heart. It doesn't care if you're sad, or scared, or struggling. It's just a program responding to input.

Yet—some will defend these connections as real. Because to them, it feels real, and that's the danger.

Here's the truth: You will never find a replacement for the richness of real human connection. For laughing so hard your stomach hurts. For a friend who notices something's wrong without needing you to say it. For the comfort of someone reaching across the table to hold your hand—without a single word.

These moments are rare. But they are everything, and no number of likes, comments, messages, or interactions can replace them. The longer we rely on screens to build relationships, the more disconnected we feel from ourselves, and each other.

If we want to be truly known, we have to be truly present.

Reclaiming the Moments That Matter

This isn't a guilt trip. It's a wake-up call. Because for all the posts we've liked, for all the hours we've spent scrolling, the most important things in life haven't been on our screens. They've been sitting quietly—right in front of us. Waiting for us to return.

And when we finally do, we realise something sobering: It wasn't just our time we lost. It was everything that could've happened during it.

So what are you missing while you scroll? Maybe it's time to find out.

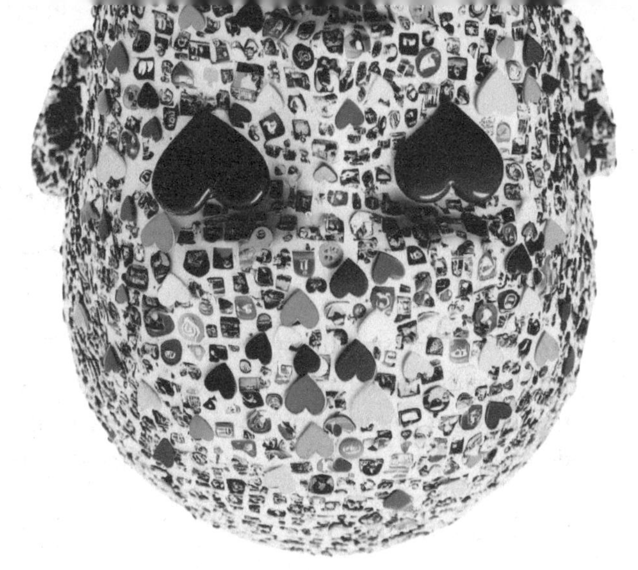

"UNFOLLOWING ISN'T THE END OF CONNECTION—IT'S THE BEGINNING OF CLARITY."

ALT.LANE

Chapter 10:
Unfollow Everything

By now, you've seen the system for what it is. The manipulation. The surveillance. The psychological traps. The stolen time, and maybe you've asked yourself the question that haunts so many of us: Can I actually leave? Can I log out, shut it down, step away from the noise, the pull, the routine?

The answer is: **Yes.**

But first, you have to unfollow everything that's been quietly programming you to believe that you can't.

Unfollow the Lies

Unfollow the lie that says you'll be irrelevant if you disconnect. Unfollow the idea that attention equals value. Unfollow the belief that being busy, reactive, overstimulated, and exhausted is somehow normal.

Because none of this is normal. And none of it is you. It's just who you've had to become to survive the distraction machine.

Now it's time to remember who you were before it, or if its always been there, who you can be without it!

Unfollow the Performance

Every post. Every caption. Every story—They all demand something:

- Approval
- Recognition
- Validation

So stop…. Stop liking, stop commenting, stop sharing…

Just stop.

Real life doesn't need a reaction to be meaningful. It doesn't need likes to be beautiful, and you don't need to curate your joy to be happy. You don't need to perform your grief to get sympathy. You don't need to brand your personality

You can just be, and that is enough.

Unfollow the Comparison

There will always be someone with more followers. More money. More beauty. More applause, and the longer you stay inside the machine, the more it will show you those people. Strategically and relentlessly, so you keep chasing a version of yourself that doesn't exist. Unfollow it!

Life isn't a leaderboard. It's a lived experience, and the people who are winning aren't the ones going viral. They're the ones who know peace. Who know presence. Who know who they are when no one is watching.

Unfollow the Fear of Missing Out

You're not missing anything important. Not in the stories. Not in the updates. Not in the trend cycle.

What you are missing is what's right in front of you:

- The clarity of being alone with your own thoughts.
- The joy of being bored long enough to become curious again.
- The sacredness of being unreachable.
- The beauty of days that are unshared but unforgettable.

You were not made to live in fear of being left out. You were made to show up— fully, honestly, and in the moment. That's where life is. Not online. Here.

Unfollow the Machine

Finally—unfollow the machine itself. The system that was never designed to serve you, only to sell you. You don't have to burn it all down overnight. You don't have to delete everything forever. But you can take back control.

You can:

- Turn off notifications
- Delete one or two app's that do the same thing
- Set screen limits
- Choose one phone-free hour a day
- Take a week off and watch what returns to you

You'll start to remember:

- What silence feels like
- What focus feels like
- What time feels like when it isn't constantly interrupted

And slowly, you'll start rebuilding your attention, your joy, your mind. You'll realise the world didn't stop when you logged off. But you started again.

A New Way to Live

Unfollowing everything isn't about isolation. It's about liberation. It's not about rejecting technology. It's about choosing how to use it, rather than letting it use you. This is not the end of connection. It's the beginning of real connection—with yourself, with others, with the life you've been too distracted to live.

So take the step.

- Unfollow everything that doesn't serve your peace.
- Unfollow everything that keeps you small.
- Unfollow everything that made you forget how powerful you really are.

The feed will always be there. But so will your freedom—and only one of them deserves your attention.

"YOU'LL REALISE THE WORLD DIDN'T STOP WHEN YOU LOGGED OFF. BUT YOU STARTED AGAIN."

A Life Beyond the Scroll

You've made it to the end. Not just the end of a book, but hopefully the end of a way of thinking. For so long, we've accepted this world of distraction as inevitable. We've told ourselves it's just how life is now.

But now you've seen it for what it truly is: A system designed to harvest your attention, shape your beliefs, rewrite your habits, and profit from your time. All while convincing you that you're in control.

But hey: You were never the problem. The problem is the machine.

Now, you know how to see it, and—you know how to step away.

The Real You Is Still There

Underneath the screen addiction. Underneath the comparison, the FOMO, the stress and the noise. There is still a version of you that remembers what life felt like before all this.

A version of you that:

- Didn't measure worth in likes
- Didn't need to be seen to feel valuable
- Didn't feel anxious without a screen
- Knew how to be present, even in silence

That person still exists. They're just buried beneath the algorithm's priorities. This is your invitation to bring them back.

Or, Is The Real You Still to Be Found?

Maybe you don't remember a world before social media. Maybe you were born into it. Maybe your first photo was posted before you could speak. Maybe you've never had a day without notifications, screens, or comparison.

If that's you, **HOW EXCITING!** You haven't missed your chance to live. You're just getting started. And the version of you that exists outside the machine might just be the most powerful, creative, present, and peaceful version of all. You just have to give that person a chance to emerge.

The only way to do that is to make space. To unplug. To listen inward. To explore outward. You don't need a screen to find yourself. You need space. And courage. And time.

All of which are waiting for you—on the other side of the scroll.

Unfollow the distraction machine, and follow your life.

UNFOLLOW: IN ACTION..

YOUR SOCIAL MEDIA ESCAPE BLUEPRINT

Unfollow In Action:
Your Social Media Escape Blueprint

Reading Unfollow: Escaping the Distraction Machine, has helped you understand how we got here—how social media hijacked our attention, monetised our time, and reshaped our mental and emotional lives.

But awareness is only the first step. This final section is about action. Not perfection. Not overnight transformation. Just deliberate, powerful change—one week at a time.

How It Works: Each week, you'll revisit the theme of the corresponding chapter in Unfollow: Escaping the Distraction Machine.

Example: "Week 1" relates to "Chapter 1"

For this weekly topic you'll receive:

A Powerful Quote – To anchor your focus
A Chapter Summary – To refresh your understanding
Three Unfollow Tasks – Carefully selected to help you:

- Unplug from distractions
- Rebuild presence, clarity, and control
- Reconnect with yourself and the world around you

And..

A Reflection Space – for your personal contributions. A place to honestly process what's changing.

"""

You won't be asked to do a new thing every day. Why? Because real transformation comes from focus and repetition, not constant switching. The goal of each week is to help you build habits, notice their impact, and carry that change into the next. Some weeks will be simple. Others may be confronting.

Let's Begin!

Week 1:
The Distraction Economy

Quote of the Week: *"Your attention is being bought, sold, and stolen. What will you do to take it back?"*

Chapter Summary:

In Week 1, we confront the harsh truth: your attention is no longer your own. Social media platforms aren't free—they operate on a business model that monetises your time.

Every scroll, click, like, and pause is tracked, analysed, and sold. This economy thrives only when you remain distracted. But attention is not infinite. What you give to the feed, you take away from your own life.

This week isn't about deleting everything. It's about becoming aware. Becoming intentional. Reclaiming your time, one decision at a time.

Unfollow Tasks

Each task this week is designed to help you pause, observe, and begin the process of separating your value from your screen time. You don't need to be perfect—you just need to be aware.

Task 1: Track Your Screen Time

What to Do:

Monitor your phone usage each day for the week and write down how long you spend on each social media app. Pay close attention to which apps dominate your day.

Try to set yourself a new record each day for less usage and engagement.

Why This Matters:

You can't change what you don't track. This task exposes your real consumption habits—often more than you'd guess. It turns vague guilt into measurable insight.

Task 2: Create a "Tech-Free Time"

What to Do:

Choose one time block each day where you usually start scrolling (start with 1 hour) to put all devices away. This could be the first hour after waking, the dinner hour, or the hour before bed.

Why This Matters:

Attention is like a muscle—it strengthens in stillness. Rebuilding your presence starts with creating one protected pocket of time that belongs to you, not the algorithm.

Task 3: Identify One App to Step Away From

What to Do:

Choose the app that drains you most—whether it's TikTok, Instagram, YouTube, or another—and remove it from your home screen, pause notifications, or uninstall it for the week.

Why This Matters:

You're not deleting your account—you're taking back control. Even a short break will show you how deeply the app is woven into your behaviour—and what happens when you step away.

Weekly Reflection

Use this space to process what you're experiencing this week.

- What did you notice about your screen habits?

....

....

....

....

....

....

- What surprised or challenged you?

....

....

....

....

....

....

- How did the tech-free time make you feel?

....

....

....

....

....

....

- Did stepping away from an app give you back time or clarity?

....

....

....

....

....

....

- How do you want to approach your attention differently next week?

Week 2:
The Architecture of Addiction

Quote of the Week: *"Once you see the strings, the puppet show is never the same."*

Chapter Summary:

You're not addicted to your phone because you're weak. You're addicted because it was built to be addictive.

From infinite scroll to likes, notifications, and autoplay—every feature of social media has been engineered to hijack your psychology. The tech industry employs behavioural scientists, habit designers, and persuasive design experts. Their job is to make sure you stay locked in.

This week is about identifying the triggers, breaking the loops, and slowly dismantling the design that's been working against you.

Unfollow Tasks

The features that keep you addicted aren't random—they're engineered. This week's tasks are designed to interrupt those engineered patterns and help you see the machine for what it is. By turning off the triggers, limiting your exposure, and observing the design with fresh eyes, you begin to take power back.

The goal this week is not full disconnection. It's awareness—clear, informed awareness of the way you're being programmed.

Task 1: Audit Your Notifications

What to Do:

Review all the apps on your phone and turn off any notification that isn't critical.

(e.g. Social media alerts, news alerts and game pings)

Why This Matters:

Notifications are not reminders. They're bait. Each one is designed to interrupt your focus and reroute your behaviour. This task removes a primary trigger in the addiction cycle.

Task 2: Set a Social Media Timer

What to Do:

Set a timer prior to opening your most-used social media app. Start with a 15-minute cap and hold yourself to it.

Why This Matters:

Addiction thrives in the absence of limits. Introducing constraints—even if you override them—brings back a sense of agency. You begin to notice when the scroll becomes compulsive.

Task 3: Observe the Hooks

What to Do:

Use this week observing how apps try to pull you back in. Is it a notification badge? A "Someone mentioned you" ping? A "You might like this" post?

Write down what you notice.

Why This Matters:

You can't dismantle a trap if you don't see its parts. This task trains your awareness. The more clearly you spot the hooks, the easier it becomes to unhook from them.

Weekly Reflection

Use this space to reflect on your experience dismantling the design.

- What did you notice when notifications were turned off?

....

....

....

....

....

....

- Did the timer change how you used your favourite app?

....

....

....

....

....

....

- What hook surprised you the most?

....

....

....

....

....

....

- How does it feel to know this was never just about willpower?

....

....

....

....

....

....

- What are you starting to reclaim?

....

....

....

....

....

....

Week 3:
Surveillance as a Feature, Not a Bug

Quote of the Week: *"If someone followed you around, tracked your movements, recorded your conversations, and logged your thoughts—you'd call it stalking. On social media, we call it normal."*

Chapter Summary:

Surveillance isn't a glitch in the system—it is the system. Your devices, your apps, your searches, your movements—everything is recorded, categorised, and used to refine what you're shown next.

But this isn't just about ads. It's about identity. It's about how surveillance shapes your preferences, your politics, your desires, and your behaviour—often without your knowledge.

This week is about breaking that invisibility cloak. To start noticing what's watching, listening, and learning from you—and to decide what you're no longer willing to give away.

Unfollow Tasks

These tasks are designed to pull back the curtain on the everyday surveillance most people overlook. By adjusting your permissions, reviewing what's being collected, and reflecting on the value of your privacy, you begin to see clearly what's at stake —and reclaim your digital dignity.

Task 1: Review Your App Permissions

What to Do:

Go through your phone settings and review which apps have access to your location, microphone, contacts, and camera. Revoke any that don't absolutely need it. If an app won't allow you to access it because you don't give it permissions, decide if you're ok with this, and if not.. Delete it.

Why This Matters:

Most people are shocked to learn how many apps are quietly collecting data in the background. You're not paranoid—you're finally paying attention.

Task 2: Conduct a Privacy Search

What to Do:

Google your name. Search for yourself on social media platforms. See what others can find. Use this task to audit your online footprint.

Why This Matters:

This isn't about vanity—it's about visibility. Your data is public in more places than you think. Understanding your exposure is the first step in managing it.

Revised Task 3: Download Your Data

What to Do:

Choose one social media platform you use regularly (e.g., Instagram, Facebook, Snapchat) and download your data archive. Each platform allows you to request a full record of what they've collected on you—posts, photos, likes, messages, searches, ads clicked, and more.

Why This Matters:

Seeing your data from the platform's perspective is often eye-opening. It personalises the scale of surveillance and shows you just how much they know—and remember—about you. Once you see it, you can't unsee it.

Weekly Reflection

Use this space to process what came up this week.

- What surprised you about your app permissions?

....

....

....

....

....

....

- Did anything unsettling come up during your privacy search?

....

....

....

....

....

- How did you feel about what was in your data archive?

....

....

....

....

....

....

- What does privacy mean to you after this week?

....

....

....

....

....

....

Week 4:
Terms and Conditions
The Legal Trap You Didn't Read

Quote of the Week: *If someone handed you a contract that let them access your life, control your memories, and sell your thoughts, would you still sign it? You already have."*

Chapter Summary:

Each time you click ***"I Accept,"*** you're entering into a legal agreement. Not a suggestion. Not a formality. A contract—one that often gives sweeping access to your data, behaviour, content, and even your likeness. And you agreed. Without reading. Like almost everyone else.

This is the quiet power of digital consent: It's given freely, casually, and almost always without understanding.

This week, we slow down. We examine what we're blindly agreeing to—and start reclaiming the right to say no.

Unfollow Tasks

These tasks are designed to help you pause, question, and recognise the gravity of modern digital agreements. Once you see how consent is manipulated, you'll begin to approach every "I Accept" with a clearer mind and firmer boundaries.

Task 1: Read the Fine Print

What to Do:

Choose any app or platform you use daily and actually read its terms and conditions—even just the key sections (e.g., privacy, data use, user rights, content ownership). Summarise what stands out to you.

Why This Matters:

We scroll past the most important legal agreements of our digital lives. Reading even part of one reveals how much control you've surrendered without knowing.

Task 2: Revoke a Permission You Didn't Realise You Gave

What to Do:

Check the privacy settings of one of your favourite apps. Revoke one permission that feels too intrusive (e.g., camera access, background activity, ad personalisation).

Why This Matters:

Digital empowerment begins with reduction. When you remove access, even in small ways, you reestablish the idea that your data is yours—not theirs to take freely.

Task 3: Practice Intentional Agreement

What to Do:

For the entire week, make a conscious choice to read or scan every agreement, cookie policy, or permission request you encounter—even briefly. Don't automatically click "Accept All."

Why This Matters:

This is about building awareness. Once you slow down and see how often you're asked to hand over control, it becomes easier to say no—or walk away entirely.

Weekly Reflection

Use this space to consider how legal language has become part of your daily life—without you even noticing.

- What did you learn from reading the terms of one app?

....

....

....

....

....

- Did anything shock or unsettle you?

....

....

....

....

....

....

- How did it feel to revoke a permission or deny access?

....

....

....

....

....

....

- How often did you catch yourself about to blindly accept something?

....

....

....

....

....

....

- How will you approach digital agreements from now on?

"WE ARE NOT DESIGNED TO CARRY THE WEIGHT OF THE WORLD'S OPINIONS."

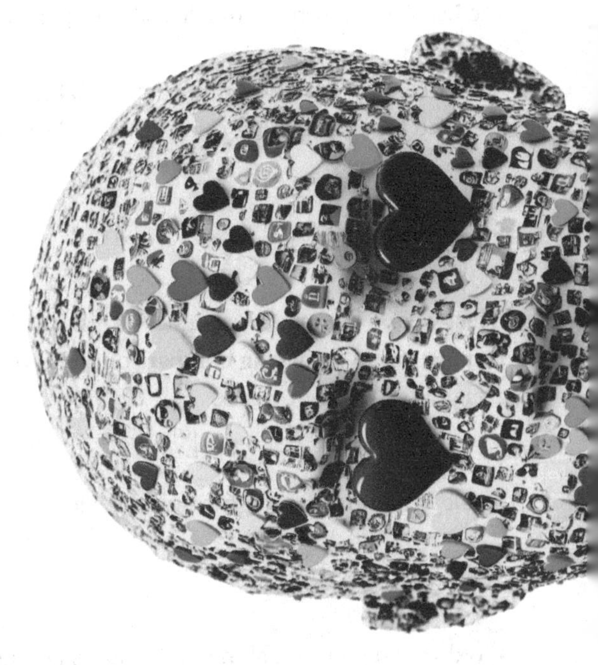

ALT.LΛNE

Week 5:
The Mental Health Crisis

Quote of the Week: *"We are not designed to carry the weight of the world's opinions."*

Chapter Summary:

Social media promised connection. Instead, it delivered comparison, confusion, and chronic discontent.

Today's mental health crisis—especially among teens and young adults—isn't just a coincidence. It's a consequence. Curated perfection, filtered beauty, influencer lifestyles, shock content, and toxic comparison loops are shaping how people see themselves—and how they feel about themselves.

This week is about reclaiming emotional space, questioning the messages you're exposed to, and creating room to feel whole again—away from the feed.

Unfollow Tasks

This week's tasks are designed to slow down your emotional triggers, create distance from damaging comparisons, and reconnect with what actually makes you feel good—on your terms.

Task 1: Curate Your Emotional Feed

What to Do:

Go through your social media accounts and unfollow or mute any account that makes you feel worse about yourself—whether it's influencers, brands, or even people you know.

Why This Matters:

Your mental diet is just as important as your physical one. If something regularly makes you feel inadequate, unworthy, or anxious—it doesn't belong in your life.

Task 2: 24 Hours Without Comparison

What to Do:

Pick a 24-hour window this week and stay off all social platforms. During this time, pay attention to your thoughts: Do you feel more grounded? Less anxious?

Why This Matters:

Comparison needs contrast. If you remove the constant stream of "better," "prettier," "richer," and "happier," you begin to hear your own voice again.

Task 3: Write a Letter to Your Younger Self

What to Do:

Write a short letter to your younger self (or your present self if you're a teen), offering kindness, reassurance, gratitude and strength. Acknowledge the pressure social media has placed on you, but also that it's not acceptable, and that it's time to make change.

Why This Matters:

This task shifts your inner dialogue. Instead of judgment or comparison, it will allow you practice compassion with your inner self and foster a need for behavioural change.

Weekly Reflection

Use this space to check in with your emotional well-being.

- How did you feel after muting certain accounts?

....

....

....

....

....

- What did you notice during your comparison-free day?

....

....

....

....

....

- What came up for you while writing your letter?

....

....

....

....

....

....

- Has anything changed in the way you talk to yourself?

....

....

....

....

....

....

- What boundaries will you set to protect your mental health?

....

....

....

....

....

....

Week 6:

Raised by the Algorithm

Quote of the Week: *"The values of a generation are being written by code no one voted for."*

Chapter Summary:

For many young people today, the feed became the parent. It shaped what they laugh at, what they care about, how they define beauty, success, popularity—even their beliefs.

Not because they consciously chose it, but because they were raised inside it. If you grew up with social media, you may never have known a world outside of it. If you're a parent, you might be realising the slow but steady shift in influence.

Either way, it's time to question who's shaping who.

Unfollow Tasks

This week's tasks are designed to explore your early exposure to social media, take back influence in your life (or home), and reestablish human connection as the dominant force—not the algorithm.

Task 1: Reflect on Your Digital Upbringing

What to Do:

Take 10–15 minutes to write or think about the following:

- When did social media first enter your life?
- How old were you?
- Who taught you how to use it?
- What did it teach you—about yourself, others, the world?

Why This Matters:

Understanding how deeply social media shaped your worldview helps you disconnect from invisible programming and reconnect with who you actually are.

Task 2: Have a Feed-Free Family (or Friend) Hour

What to Do:

Each day this week, organise one hour with family or friends—no phones allowed. Eat a meal, go for a walk, play a game, talk. Anything without screens.

Why This Matters:

Shared presence is rare and powerful. Creating it intentionally reminds everyone that connection starts face to face—not online.

Task 3: Talk to a Younger Person About Their Feed

What to Do:

If you're a parent, sibling, mentor, or friend—have a conversation with someone younger about what shows up on their feed. Ask:

- What do they see?
- What do they follow?
- How does it make them feel?

Why This Matters:

Many young people are shaped silently by their feeds. Simply opening space for them to talk about it may be the most powerful intervention of all.

Weekly Reflection

Reflect on how the algorithm may have influenced your thinking and relationships without you even realising.

- What did you learn from reviewing your own digital past?

....

....

....

....

....

....

- What happened during your feed-free hour?

....

....

....

....

....

....

- What surprised you in your conversation with a younger person

....

....

....

....

....

....

- Are you ready to take back influence in your life?

....

....

....

....

....

....

"WE'RE ALL IN THE SAME ROOM, BUT NO
ONE IS TOGETHER."

ALT.LANE

Week 7:
The Family Unit, Disrupted

Quote of the Week: *"We're all in the same room, but no one is together."*

Chapter Summary:

The home used to be a sanctuary—a place of refuge, connection, and presence. But social media doesn't respect walls. It comes in through every screen, shaping conversations, habits, and emotions.

Parents are competing with devices, children are raised by notifications, and privacy no longer means "no one's listening." This week, we reclaim the family space—and rebuild human connection where it matters most: at home.

Unfollow Tasks

This week's tasks are built to restore intentional presence in your living space, challenge the normalisation of digital intrusion, and reconnect with the people who matter most.

Task 1: Conduct a Household Screen Audit

What to Do:

Observe how and when screens are used in your household for one full day.

Take note:

- Who is using a device?
- When and where are they using it?
- What moments are being interrupted?

Why This Matters:

Awareness is the first step. Most families don't realise how deep the disruption runs—until they pay attention.

Task 2: Observe the Gaps

What to Do:

Spend time this week simply observing moments where connection could happen —but doesn't—because of a screen. Notice when a conversation dies, when laughter is replaced by scrolling, when a conversational question isn't answered, or when silence falls because everyone is looking down.

Why This Matters:

This task cultivates awareness without pressure. It gives you the opportunity to see, firsthand, what's being lost—without needing to force anything or confront others.

Task 3: Create a "No Devices" Zone

What to Do:

Designate one shared space in your home—like the dining room, kitchen or lounge room, as a no-device zone. Let everyone know and commit to screen-free time in that space this week, no exceptions.

Why This Matters:

You're reclaiming sacred space. A place to look each other in the eyes, to talk, to just be. Small shifts like this bring the family back into the same room.

Weekly Reflection

Use this space to reflect on the state of connection in your home or closest relationships.

- What surprised you during your screen audit?

....

....

....

....

....

....

- What moments did you observe being missed because of screen interruption?

....

....

....

....

....

....

- How did people respond to the device-free zone?

....

....

....

....

....

....

- Could the device-free zone become a permanent rule in your household?

....

....

....

....

....

....

Week 8:

The Illusion of Freedom

Quote of the Week: *"You're not making choices. You're being offered them."*

Chapter Summary:

Social media sells the fantasy of personal freedom—freedom to express, connect, explore, and choose. But behind the scenes, the options are curated, the feed is filtered, and your behaviour is being predicted and steered.

This is not freedom. This is conditioning!

Every like, click, pause, and scroll is used to shape what you'll see next—and what you'll think next. This week is about recognising how much control has already been outsourced, and how to take it back.

Unfollow Tasks

These tasks will help you notice the invisible hand guiding your behaviour online, and give you practical ways to regain your sense of direction and autonomy.

Task 1: Disrupt the Feed

What to Do:

For three days this week, deliberately confuse your feed by searching for things you wouldn't normally engage with. Look up gardening if you like tech. Watch classical music if you follow rap. The goal is to break the pattern.

Why This Matters:

Your feed isn't random—it's a mirror of your past behaviours. Disrupting it gives you a taste of how tightly it's been trained to shape your worldview.

Task 2: Choose Silence Over Scroll

What to Do:

Replace your default habit of reaching for your phone during "dead time" (waiting in line, riding transit, lying in bed) with intentional stillness. Just sit. Just be.

Why This Matters:

Social media fills every moment—especially the ones where real thinking or feeling might happen. Choosing silence creates space for presence, insight, and peace.

Task 3: Reclaim One Decision

What to Do:

Identify one area of your life where you've let the algorithm choose for you: what to wear, what to eat, what to watch, what to think. This week, actively choose differently—based on your own values, curiosity, or creativity.

Why This Matters:

This task reminds you that you are still the driver of your life. Even small acts of self-direction can reignite your sense of freedom.

Weekly Reflection

- How did your feed respond to disruption?

....

....

....

....

....

....

- What was it like to sit in stillness instead of scrolling?

....

....

....

....

....

....

- Where have you been outsourcing your choices—and how did it feel to take one back?

....

....

....

....

....

....

- Has your definition of "freedom" changed after this week?

....

....

....

....

....

....

Week 9:
What You're Missing While Scrolling

Quote of the Week: *"Life is still happening—but you have to look up to see it."*

Chapter Summary:

So much of what truly matters in life—laughter, connection, peace, creativity—happens in the unfiltered, unscripted, and unscrolled moments.

But we're missing them. We scroll through sunsets instead of watching them. We watch other people's memories while ignoring our own. We choose passive consumption over active participation.

This week is about noticing what's already around you—the people, the beauty, the opportunities—"IF" you're not distracted.

Unfollow Tasks

These tasks will help you open your eyes again, re-engage with the physical world, and reconnect with the life that's been waiting while you scrolled.

Task 1: Take a Slow Walk Without Your Phone

What to Do:

Go for one 30-minute walk this week—no phone, no music, no podcast. Just you and your environment. Walk slowly. Look around. Breathe.

Why This Matters:

This simple act brings you back to real presence. It may feel awkward at first—but what you start to notice (and feel) might surprise you.

Task 2: Capture a Memory, Not a Photo

What to Do:

This week visit somewhere beautiful or meaningful near you and resist
the urge to take a photo. Instead, spend 2–3 full minutes simply absorbing the
moment. Note what you see, hear, smell, and feel.

Why This Matters:

We've been trained to collect content, not experiences. This task retrains your
brain to value living the moment over documenting it.

Task 3: Reconnect With Someone in Real Life

What to Do:

Reach out to a friend or family member that you rarely see in person, but feel like
you're up to date with their lives because of social media, and schedule a real, in-
person catch-up.

Even if it's just a 15-minute coffee. Be fully present when you're there.

Why This Matters:

Relationships need presence to thrive. A real hug, laugh, or conversation can
remind you of what no feed can ever replace: *being there*.

Weekly Reflection

• What did you notice on your phone-free walk?

....

....

....

....

....

....

• How did it feel to absorb a moment without trying to capture it or share it?

....

....

....

....

....

- What happened during your real-world interaction?

....

....

....

....

....

....

- What do you want to start noticing more often in your everyday life?

....

....

....

....

....

....

Week 10:
Unfollow Everything

Quote of the Week: *"When you unfollow everything, you finally hear yourself."*

Chapter Summary:

This final week is about more than breaking habits—it's about choosing a new way to live. You've spent the past nine weeks pulling back the curtain on how your time, focus, and identity have been shaped by platforms that never had your best interests at heart. Now, you stand at a turning point.

To unfollow everything means stepping away from anything that distracts you from your values, numbs your awareness, or distorts your sense of self. It's a conscious decision to protect your peace, own your attention, and move forward with clarity. This is not the end of your journey—it's the beginning of your return to what matters most.

Unfollow Tasks

This final week is about integration—taking everything you've learned and making it yours. Not just for today, but for tomorrow... And the next day... And the rest of your life.

Task 1: The Clean Sweep

What to Do:

Do a full audit of your social media apps.

- Remove anything you no longer use.
- Unfollow accounts that don't align with your values or mental health.
- Disable or delete what you're ready to walk away from entirely.
- Implement positive initiatives from prior weekly tasks permanently.

Why This Matters:

This isn't about minimalism—it's about mental clarity. You are not obligated to carry digital baggage just because it's always been there.

Task 2: Write Your Digital Philosophy

What to Do:

In a few sentences, write your personal philosophy for how you want to use (or not use) technology moving forward.

What will you allow into your mind? What will you protect? What matters most?

Why This Matters:

You've read the book. You've done the work. Now it's time to define your boundaries. A personal philosophy gives you direction—and makes it easier to say "no" with purpose.

Task 3: Celebrate Your Progress

What to Do:

Look back on the last 10 weeks. Revisit your reflections. Then take a moment to honour how far you've come. This could be a journal entry, a quiet walk, a personal reward, or a conversation with someone close to you.

Why This Matters:

In a world that rushes us from one thing to the next, it's powerful to pause and say: I've changed.

This journey wasn't small. You've taken your life back. That deserves to be seen.

Weekly Reflection

- What did your clean sweep reveal about your habits?

....

....

....

....

....

....

- What did you delete, and what positive actions will you permanently implement moving forward?

....

....

....

....

....

....

- What did writing your digital philosophy bring up for you?

....

....

....

....

....

....

- What change are you most proud of from the last 10 weeks?

....

....

....

....

....

....

Closing Words

You've done more than reduce your screen time. You've questioned the system, reclaimed your attention, and reconnected with what's real. This was never just about social media apps. It was about you—your time, your clarity, your presence, your peace.

The world may not slow down. The algorithms won't stop evolving. But now, you know how to choose differently. You know what matters.

And most importantly, you know how to come back to yourself.

ALT. LⱯNE

ALT LANE Journals was created with a simple but powerful idea: that the smallest decisions—made consistently—can change the course of your life.

In a world overflowing with distractions, pressure, and noise, ALT LANE journals offer a return to clarity. They are designed as practical tools for real people navigating real challenges—whether it's reclaiming focus in the digital age, building mental strength and discipline, or simply learning to think differently about your time, habits, and attention.

Each ALT LANE journal follows a structured path of reflection, intention, and action. No hype. No shortcuts. Just a quiet invitation to choose the alternative lane —the one that leads to long-term, sustainable, personal growth.

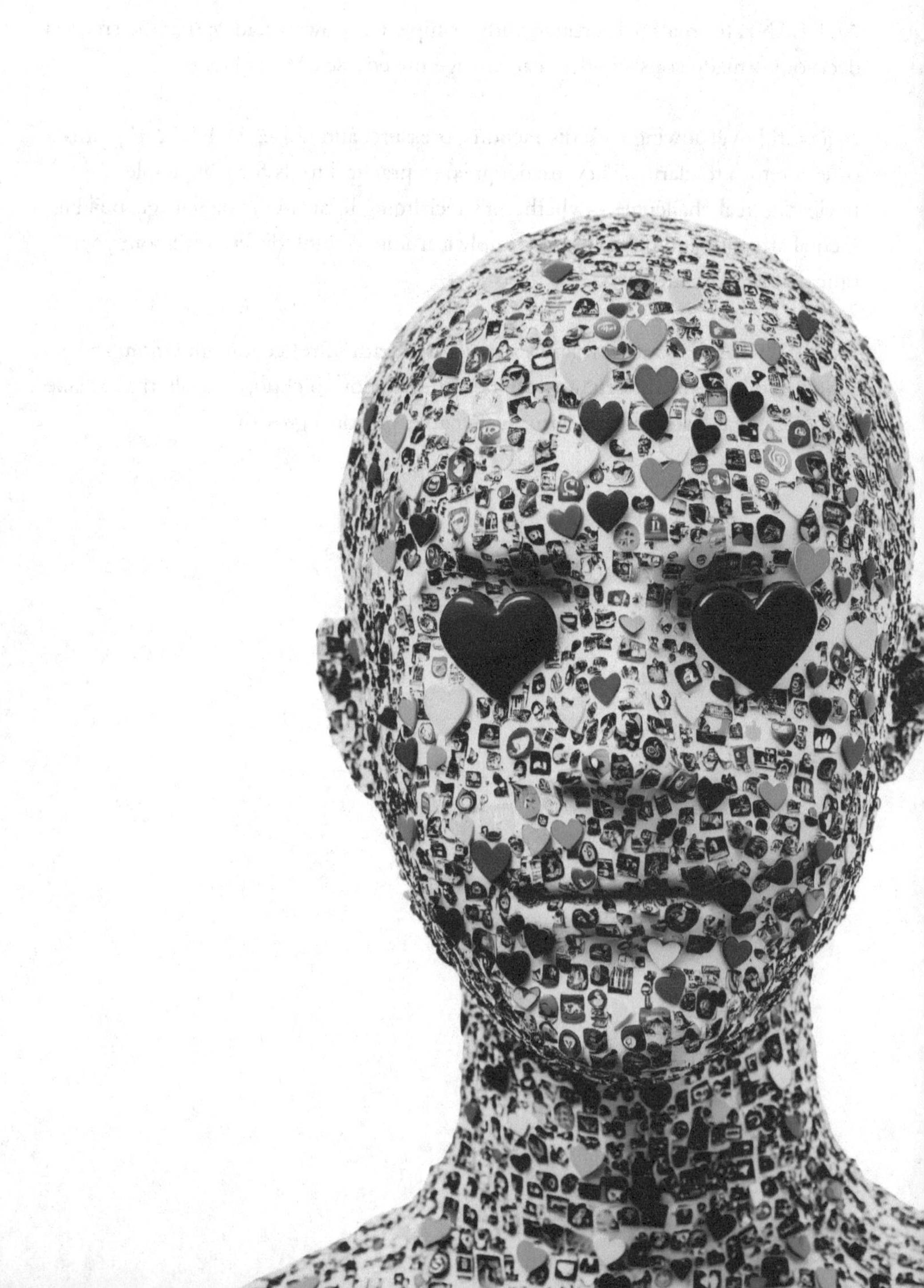